Smart Technologies for Smart Buildings

This comprehensive guide explores the integration of information systems and applications in smart buildings, emphasizing their crucial role in achieving a low-carbon economy through optimal energy efficiency and cost reduction.

This book provides cutting-edge insights into smart energy planning, including tidal energy forecasting using machine learning algorithms like XGBoost and Long Short-Term Memory (LSTM), and Internet of Things (IoT)-driven weather monitoring systems for urban smart buildings. Key features include practical implementations of serverless file management systems, Programmable Logic Controller (PLC) automation for Industry 4.0 applications, and Field-Programmable Gate Array (FPGA)-based solar-powered Electric Vehicle (EV) charging stations supporting sustainable infrastructure. Readers will benefit from the comprehensive coverage of cognitive security systems, GPS-integrated building automation, and ESP32-powered smart home automation. The publication addresses critical security considerations through vulnerability analysis in smart meters, ensuring robust implementations. These technologies enable significant energy efficiency improvements, operational cost savings, and enhanced property values while supporting sustainable environmental practices.

This publication serves developing and third-world nations where smart building technology is evolving, making it particularly relevant for universities, engineers, architects, and real estate entities implementing advanced building automation systems. The primary audience includes engineering professionals, construction organizations, property managers, and IT specialists requiring practical guidance for creating outstanding customer experiences while achieving substantial energy efficiency gains. This book also appeals to academic researchers, graduate students, and industry professionals focused on sustainable building design and smart city development initiatives.

Smart Technologies for Smart Buildings

Edited by
O. V. Gnana Swathika and K. Karthikeyan

CRC Press
Taylor & Francis Group
Boca Raton London New York

CRC Press is an imprint of the
Taylor & Francis Group, an **informa** business

Designed cover image: Shutterstock

First edition published 2027
by CRC Press
4 Park Square, Milton Park, Abingdon, Oxon, OX14 4RN

and by CRC Press
2385 NW Executive Center Drive, Suite 320, Boca Raton FL 33431

British Library Cataloguing-in-Publication Data
A catalogue record for this book is available from the British Library

Library of Congress Cataloging-in-Publication Data
Names: Swathika, O. V. Gnana editor | Karthikeyan, K. editor
Title: Smart technologies for smart buildings / edited by O.V. Gnana Swathika and K. Karthikeyan.
Description: First edition. | Abingdon, Oxon ; Boca Raton FL : CRC Press, 2027. |
Includes bibliographical references and index.
Identifiers: LCCN 2026011573 (print) | LCCN 2026011574 (ebook) |
ISBN 9781041093626 hardback | ISBN 9781041093633 paperback | ISBN 9781003649700 ebook
Subjects: LCSH: Intelligent buildings | Smart materials in architecture | Architecture and energy conservation
Classification: LCC TH6012 .S625 2026 (print) | LCC TH6012 (ebook)
LC record available at https://lccn.loc.gov/2026011573
LC ebook record available at https://lccn.loc.gov/2026011574

ISBN: 9781041093626 (hbk)
ISBN: 9781041093633 (pbk)
ISBN: 9781003649700 (ebk)

DOI: 10.1201/9781003649700

Typeset in Times
by codeMantra

Contents

Preface

Together, the suggested projects demonstrate a full ecosystem of intelligent technologies for the next generation of smart infrastructure and buildings. Tidal energy forecasting utilizing XGBoost and Long Short-Term Memory (LSTM) for coastal structures and Field-Programmable Gate Array (FPGA)-based solar-powered Electric Vehicle (EV) charging stations are examples of advanced energy intelligence that support sustainable energy planning and renewable integration. Scalable, safe, and effective data handling is guaranteed by serverless and cloud-enabled architectures like AWS S3 and Lambda-based file management systems. In Indian smart city contexts, predictive building management is further improved by Internet of Things (IoT)-driven machine learning models for urban weather monitoring. Programmable Logic Controller (PLC) systems with an Industry 4.0 focus handle industrial automation, while healthcare-focused smart management solutions—like online oxygen cylinder monitoring—extend smart infrastructure into vital services. This book aims to provide a forum for researchers and practitioners to exchange ideas and achieve progress in smart building applications, advances, and research challenges. It is highly recommended to be used as a reference book for graduate and postgraduate level university programs. Case studies range across all the chapters.

SALIENT FEATURES:

1. Intelligent Renewable Energy Integration in Smart Building
2. Cloud-Based and Serverless Architecture
3. IoT-Driven Predictive Analytics
4. Industry 4.0 and Embedded Automation
5. Enhanced Security and Cyber Resilience

About the Editors

O. V. Gnana Swathika (Member '11–Senior Member '20, IEEE, MIET) earned a BE degree in Electrical and Electronics Engineering from the Madras University, Chennai, Tamil Nadu, India, in 2000; an MS in Electrical Engineering at Wayne State University, Detroit, MI, USA, in 2004; and a PhD in Electrical Engineering at VIT University, Chennai, Tamil Nadu, India, in 2017. She completed her postdoctoral research at the University of Moratuwa, Sri Lanka, in 2019. She is currently working as a Professor and Deputy Director at the Centre for Smart Grid Technologies, Vellore Institute of Technology, Chennai, India. Her current research interests include microgrid protection, power system optimization, embedded systems, and photovoltaic systems.

K. Karthikeyan is an electrical and electronics engineering graduate with a master's in Business Administration from the University of Madras. With two decades of extensive experience in electrical design, he has contributed significantly to the building services sector comprising airports, Information Technology Office Space (ITOS), tall statues, railway stations/depots, hospitals, educational institutional buildings, residential buildings, hotels, steel plants, and automobile plants in India and abroad (Sri Lanka, Dubai, and the UK). Currently, he is Chief Engineering Manager - Electrical Designs for Larsen & Toubro (L&T) Construction, an Indian multinational Engineering Procurement Construction (EPC) contracting company. He has also worked at Voltas, ABB, and Apex Knowledge Technology Private Limited. His primary role involves the preparation and review of complete electrical system designs up to 110 kV at the detailed engineering stage, covering various electrical design calculations, design basis reports, construction-ready drawings, as well as Mechanical Electrical Plumbing (MEP) design coordination. He is the point of contact for both client and internal project teams, and also leads and manages a team of design and divisional personnel. His responsibilities include day-to-day interaction with clients, peer review, managing project deadlines, project time estimation, assisting in staff appraisals, training, and recruiting.

Contributors

V. Anantha krishnan
School of Electrical Engineering
Vellore Institute of Technology
Chennai, India

Kollipara Anirudh
School of Electrical Engineering
Vellore Institute of Technology
Chennai, India

B. Kiran Aravind
School of Computer Science and Engineering
Vellore Institute of Technology
Chennai, India

J. Muthu Bharath
School of Electrical Engineering
Vellore Institute of Technology
Chennai, India

Daksha Raghunath Bordekar
School of Electrical Engineering
Vellore Institute of Technology
Chennai, India

N. T. Bragadeeshwaran
School of Electrical Engineering
Vellore Institute of Technology
Chennai, India

G. Surya Deepan
School of Electrical Engineering
Vellore Institute of Technology
Chennai, India

Mridulay Dixit
School of Electrical Engineering
Vellore Institute of Technology
Chennai, India

K. T. M. Udayanga Hemapala
Department of Electrical Engineering
University of Moratuwa
Moratuwa, Sri Lanka

Amit Hota
School of Electrical Engineering
Vellore Institute of Technology
Chennai, India

G. Jagadish
School of Electrical Engineering
Vellore Institute of Technology
Chennai, India

R. Lalitha Kala
School of Electrical Engineering
Vellore Institute of Technology
Chennai, India

K. Kartheesan
School of Electrical Engineering
Vellore Institute of Technology
Chennai, India

R. Guru Karthik
School of Electronics Engineering
Vellore Institute of Technology
Chennai, India

Deepa Rakesh Kumar
School of Electrical Engineering
Vellore Institute of Technology
Chennai, India

Bhupesh Singh Kushwaha
School of Electrical Engineering
Vellore Institute of Technology
Chennai, India

Chenchu Lekhana
School of Electrical Engineering
Vellore Institute of Technology
Chennai, India

Maryamah
Universitas Airlangga
Surabhaya, Indonesia

Adarsh Mohapatra
School of Electrical Engineering
Vellore Institute of Technology
Chennai, India

M. S. Nanditha
School of Electrical Engineering
Vellore Institute of Technology
Chennai, India

S. N. Nadhish
School of Computer Science and Engineering
Vellore Institute of Technology
Chennai, India

Bandi Nagavindhya
School of Electrical Engineering
Vellore Institute of Technology
Chennai, India

S. Nakul
School of Electrical Engineering
Vellore Institute of Technology
Chennai, India

P. Pavithran
School of Electrical Engineering
Vellore Institute of Technology
Chennai, India

R. Pradeesh
School of Electrical Engineering
Vellore Institute of Technology
Chennai, India

S. K. Preetham
School of Electrical Engineering
Vellore Institute of Technology
Chennai, India

Andrila Priyam
School of Electrical Engineering
Vellore Institute of Technology
Chennai, India

Harish Raaghav
School of Electrical Engineering
Vellore Institute of Technology
Chennai, India

S. S. Radhikaa
School of Electrical Engineering
Vellore Institute of Technology
Chennai, India

Utkarsh Rajesh
School of Electrical Engineering
Vellore Institute of Technology
Chennai, India

A. Sachin
School of Computer Science and Engineering
Vellore Institute of Technology
Chennai, India

Kumar Sairaj
School of Electrical Engineering
Vellore Institute of Technology
Chennai, India

Tharaga Sharmilan
University of Kealniya
Kealniya, Sri Lanka

Dishank Singh
School of Electrical Engineering
Vellore Institute of Technology
Chennai, India

Ramachandran Sitharthan
Centre for Smart Grid Technologies
Vellore Institute of Technology
Chennai, India

O. V. Gnana Swathika
Centre for Smart Grid Technologies
Vellore Institute of Technology
Chennai, India

S. Subhiksha
School of Electrical Engineering
Vellore Institute of Technology
Chennai, India

Tanya
School of Electrical Engineering
Vellore Institute of Technology
Chennai, India

Smart Energy Planning in Coastal Buildings: *Tidal Energy Forecasting Using XGBoost and LSTM*

1

A. Sachin, B. Kiran Aravind, S. N. Nadhish, Maryamah, and O. V. Gnana Swathika

1.1 INTRODUCTION

Seaside smart buildings, ranging from oceanic research centers and coastal command centers to island environmental hotels, face difficult energy management due to their remote location, fluctuating weather conditions, and varying availability from the grid center. Such environments need energy systems that, in addition to being robust and autonomous, are also highly efficient and forecastable in operation.

Historically, energy systems in buildings have been operating in reactive mode, responding after the fact to a peak in demand or lack of generation following its occurrence. However, as the levels of integration of renewable sources with cyclical or intermittent characteristics increase, predictive control of energy must emerge as a change. Predictive control enables anticipatory scheduling of power-intensive tasks, dynamic load balancing, and greater utilization of local energy sources, ultimately promoting greater operational efficiency and sustainability.

Among renewable sources, tidal power has one specific benefit regarding coastal facilities. Unlike solar or wind power – both of which are extremely weather-dependent – tidal currents are governed by sound celestial mechanics. This built-in periodicity has the advantage of predicting energy generation with a great deal of precision days or even weeks in advance.

DOI: 10.1201/9781003649700-1

In order to fully exploit tidal power, accurate prediction of tidal current velocity is of prime importance. Small errors in speed predictions can lead to very significant differences in power production, given the cubic relationship of water speed to turbine power output. Near-real-time or real-time prediction allows smart building energy management systems (BEMS) to adjust battery storage operations in advance, shift Heating, Ventilation, and Air Conditioning (HVAC) schedules, or change appliance loads based on predicted energy supply.

This chapter explains machine learning (ML) applications, such as Extreme Gradient Boosting (XGBoost) and Long Short-Term Memory (LSTM) networks, for short-term tidal current forecasting. The goal is not only to evaluate predictive performance but also model interpretability, inference rates, and implementability in low-power edge computing settings common to intelligent coastal buildings.

1.2 TIDAL ENERGY AND ITS ROLE IN SMART INFRASTRUCTURE

Tidal energy, produced by cyclical rise and fall in sea level and ocean currents, is the most consistent and predictable of all renewable energy sources. Unlike solar or wind power, which change with weather patterns and time of day, tidal flows are driven by gravitational forces between the Earth, Sun, and Moon. These cosmic mechanics produce regular, semidiurnal or diurnal tidal currents that are highly predictable, and tidal power is a strong contender for supply to smart energy systems that require it to be guaranteed.

The fundamental principle in tidal energy generation lies in harnessing the kinetic energy of moving water. While flowing through marine turbines – either ebb or flood tides – water drives rotors on generators, converting mechanical energy to electrical energy. The developed energy is a function of the cube of the velocity of the current, illustrating the huge need for exact forecasting and real-time observation of water velocity in order to achieve optimal output.

In smart infrastructure parlance, tidal energy stands out for its advantages. Coastal zones featuring marine research institutions, water resorts, or residential colonies grow to require energy independence due to their limited access to secure grid connectivity. Tidal energy systems can be installed in situ through the use of microturbines or floating platforms strategically located at estuaries, bays, or offshore channels. They can serve as the primary or supplementary power source of in situ structures, offering a clean, silent, and constant flow of power with minimal occupation of land.

Integrating tidal power into coastal smart buildings opens the way to more advanced energy management techniques. Smart houses can prearrange high-power appliances in advance at times of maximum tidal power generation. Experiential cycles or data processing loads can be coordinated by research facilities based on prevailing power supply expectations. Resorts may coordinate lighting, heating, ventilation, and water heating functions with the tidal cycle, reducing the cost of operation while fueling sustainability.

Yet to realize these gains in full, forecasting tidal currents with precision and timeliness is critical. This is where data-driven advanced methods such as ML come into play, allowing high-frequency forecasts that facilitate real-time optimization and smart automation up and down the smart infrastructure stack.

1.3 LITERATURE LANDSCAPE

Accurate prediction of renewable resources has generally been a major enabler of smart grid operations. Prediction enables systems to plan energy delivery, plan loads for optimization, and make informed decisions on battery use and storage, while maintaining stable grids. In the case of coastal buildings that are looking at integrating tidal power, this challenge is particularly severe, in that it requires coordination of power production with building usage patterns in changing maritime environments.

Traditionally, tidal forecasting has relied on physics-based models rooted in hydrodynamic simulation. Such methods replicate ocean current behavior according to bathymetry, tidal constituents, and complex fluid dynamic equations. While physics-based, the models are computationally expensive and possess high-resolution environmental inputs. Pretuned parameters upon which they rely diminish their capacity for dealing with localized inaccuracies or unanticipated events. Additionally, these models are difficult to scale to low-power real-time applications in embedded systems.

Statistical forecasting approaches such as autoregressive models (ARIMA), Kalman filters, and time-series decomposition have been employed in marine energy systems to overcome these limitations. These models are computationally efficient with faster runtime but do not capture the nonlinear, multi-scale temporal behavior inherent in tidal flows, especially in noisy, real-world data regimes.

ML has emerged as a serious contender in the last decade, particularly for renewable energy forecasting. In solar and wind power systems, ML models including decision trees, support vector machines (SVMs), random forests, and deep learning architectures have seen widespread applications. Such models have the capability to learn complex patterns from past data directly without having to encode physical laws explicitly.

In tidal energy, however, ML remains to be fully utilized. Some recent works applied recurrent neural networks (RNNs), including LSTM networks, to predict tidal velocities with promising outcomes. Ensemble models such as XGBoost are also revealed to be extremely powerful in comparable time-series tasks. The majority of research, however, is lacking with regard to accuracy comparisons in desktop environments, without mentioning deployment viability.

This highlights a critical research gap: existing ML models for tidal forecasting are rarely experimented on deployment on edge devices or within smart grid control loops. In real-world smart infrastructure, models must be light, fast, and interpretable. Black-box models are precise but are oftentimes suspicious in operational use since they are unclear. In addition, most prior work does not use a standard pipeline or shared dataset, so comparisons among models are challenging.

Thus, an integrated forecasting approach is required that has the best prediction performance and real-time deployability, interpretability, and low computational overhead, especially in embedded coastal environments. This need is met in this chapter by comparative investigation of XGBoost and LSTM models with the same experimental pipeline, with explicit emphasis on deployment-based measures.

1.4 FORECASTING FRAMEWORK FOR COASTAL ENERGY PLANNING

As more coastal smart buildings source their energy from intermittent renewable sources like tidal energy, the industry requires more sophisticated energy planning that can respond to predicted availability, rather than past consumption. In order to facilitate the shift, a robust forecasting framework must exist, one that not only provides good short-term tidal current forecasts but also integrates smoothly into BEMS. This section presents the objectives and architectural design of such a forecasting framework, specifically tailored for real-time deployment over coastal infrastructure.

1.4.1 Objectives

The main objective of this forecasting framework is to enable coastal buildings to forecast and align their energy consumption with tidal energy availability. This involves three main goals:

Predict Tidal Current Speeds and Power Output: By means of ML models trained on historical or synthetic data, the system attempts to make sound predictions of momentary velocities. These are subsequently converted into estimates of power output by physical equations that convert current speed to turbine energy output. This allows buildings to predict energy availability in advance.

Allow Energy Control Decisions: With predicted power output data, a building's energy management system can prioritize and delay tasks, offset loads, and optimize energy storage. For example, desalination facilities or HVAC units can be operated when predicted tidal highs occur, and non-essential loads shifted to low-demand periods.

Evaluate Feasibility of XGBoost and LSTM in Building Energy Controllers: In addition to accuracy, the models are also verified for inference time, memory footprint, model size, and interpretability. This is to validate that the selected forecasting approach is hardware- and real-time-compliant with the conventional embedded controllers used in smart buildings.

1.4.2 Conceptual Architecture

The forecasting framework is set up as a modular system with prediction capabilities integrated into the building energy infrastructure. The structure consists of four fundamental layers:

Sensor Layer: Tidal current sensors (e.g., bottom-mounted flow meters or Acoustic Doppler Current Profilers (ADCPs)) continuously monitor current water velocity near the structure or platform.

Preprocessing and Feature Engineering Layer: Data come as unstructured inputs, which are transformed into structured inputs by using time-based encodings (e.g., hour-of-day sine/cosine), lag features, and normalization. Preprocessing is optimized for low-latency edge environments.

Forecasting Engine: A lightweight ML model XGBoost or LSTM is deployed on a local controller or edge device. It consumes the engineered features and generates short-term predictions for tidal current speed and associated power output.

Energy Control and Scheduling Layer: The projected availability of energy is used to schedule appliances for optimization, battery charging/discharging, and load shedding policies using the building's energy controller (e.g., BEMS or EMS).

The entire system is designed to operate autonomously with minimal human intervention through a control-feedback loop that responds to new sensor data. Figure 1.1 illustrates this end-to-end architecture, focusing on data flow from tidal sensors through modules of prediction and into intelligent control systems eventually.

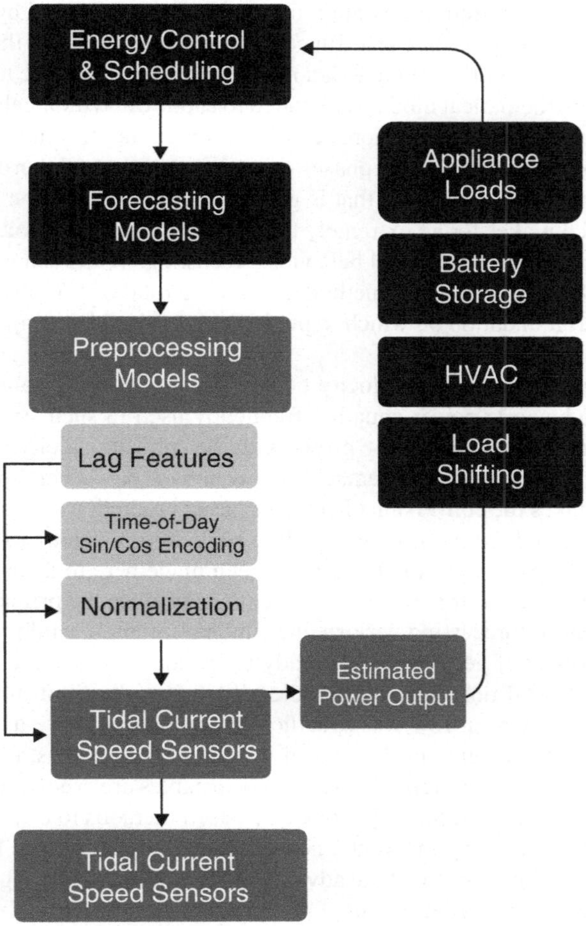

FIGURE 1.1 End-to-end architecture of smart coastal building energy management with integrated tidal forecasting layer.

1.5 DATA SIMULATION AND FEATURE ENGINEERING

Ease of access to high-resolution tidal currents data in a consistent and reliable manner remains a major challenge to engineers and scientists committed to the task of tapping renewable sources of energy along the coast. To overcome the limitation under present circumstances in an effective way, a synthetic dataset was generated with the objective of closely mimicking the natural pattern of tidal movement, at the same time keeping room for flexible experimentation and reproducibility in different situations. Utilizing synthetic data in the new approach ensures that key parameters like noise levels, sampling intervals, and number of cycles are under control, which together make it a suitable candidate for the benchmarking of ML models. The approach also significantly enhances scalable testing under edge-deployment circumstances, where local sensors can be sparse, unreliable, and intermittently unavailable for taking accurate measurements.

The dataset that was generated in preparation for analysis was produced by a well-designed simulation process employing a mixture of a sine function to accurately model the cyclical and periodic nature of tidal flows, with the addition of Gaussian noise, which adds realistic and authentic variability to the data. The specific mathematical model is intended to capture and duplicate both the deterministic and stochastic aspects inherent to tidal motion, accurately simulating the environmental variation that occurs as a result of factors like wind shear, phases of the lunar cycle, and thermal stratification effects on bodies of water. Every single data point that is generated in the simulation directly corresponds to a 5-minute interval, which makes for an extremely fine-grained temporal resolution that is particularly well-suited to practical use in real-time tidal behavior forecasting. When using the simulation over the course of an extended 48-hour period, the method generates a total of 576 observations, which makes for a solid and substantial foundation on which to perform supervised learning exercises in data analysis and prediction.

Apart from considering the present velocity of the system, the power output was also estimated using a simplified hydrodynamic power equation frequently used in such studies. It must be emphasized here that kinetic energy in fluid motion grows with the cube of velocity, and therefore the power output (P) is estimated by the relation mathematically given by $P \propto V^3$. This specific relation is highly convenient to use, as it makes the conversion of velocity-based predictions to more relevant energy estimates possible, which is of high utility in smart building applications where load scheduling and energy optimization are of high importance. An attempt has been made here to keep the study consistent by assuming constant values for the turbine area and water density since the primary interest of this study is comparative assessment of forecasting performance and not accurate calibration of the turbine.

A key and critical aspect of getting models ready is revealed through the application of effective feature engineering. With XGBoost, a strong gradient boosting algorithm that excels on structured tabular data, every individual input row was specifically designed to serve as a snapshot that retains relevant temporal dynamics. Features in this model included lag variables, i.e., capturing velocity at the previous one, two, and three timesteps. These lagged variables are crucial for helping capture autocorrelated behavior in the signal and thereby allow tree-based methods to effectively pick up temporal patterns, despite these methods being inherently non-sequential in nature. The LSTM model, however, specifically designed for applications that take advantage of sequential learning, was trained on sliding windows that span 12 consecutive observations. This particular structure enables the LSTM network to retain hidden memory states that continuously change with time, which makes it especially robust and efficient at predicting time series with long-term dependencies.

In order to allow the models to leverage the time-of-day effect that is intrinsic to tidal behavior, cyclic time features were constructed by performing sine and cosine transformations of the hour field. These representations are continuous between 23:59 and 00:00, which otherwise is not maintained when using raw hour values. This allows the models to generalize across daily cycles and enhance identification of periodic tidal trends. Additionally, all the features were normalized using Min-Max scaling to avoid domination by high-magnitude variables and enhance convergence during training. This is especially important in models such as LSTM, where stability of gradients is essential in learning long sequences.

Along with greatly improving the compatibility of many different disparate models, this well-designed feature engineering process has a real and considerable effect on the overall generalizability and robustness of the overall forecasting system as a whole. Given that the downstream application is in integration into complex smart building control cycles, it is crucial that the models not merely be able to perform but also be well-suited to produce stable and consistent predictions in a large variety of different scenarios. The judicious selection of lagged variables works to effectively buffer the model from short-term fluctuations that can occur in the data, and the use of cyclic encodings works to allow the model to balance its predictive biases with the natural increase and decrease of tidal cycles. Combined, these well-thought-out designed inputs serve to play an important part in calibrating both the predictiveness and operational reliability of the entire system, and in doing so, making it reliable in real-world applications.

Another important consideration that must be taken into account is the frequency and resolution of the used dataset. The sampling period of 5 minutes was selected with caution and consideration: it is an optimal and effective trade-off between the need to capture sufficient temporal granularity for real-time control use cases and the need to restrict computational overhead for effective deployment at the edge. Most energy controllers for intelligent buildings act over decision windows that are commonly between 5 and 15 minutes, and therefore this specific time scale is extremely well-suited to effective deployment in such systems. Furthermore, in the event of multiple sampling frequencies being necessary for certain use cases or scenarios, the synthetic dataset may be downsampled or recreated accordingly, clearly showcasing its unmatched flexibility and usability in addressing multi-faceted needs.

The simulation also incorporates artificial anomalies and drift events, which are introduced strategically to make the models under consideration robustly tested. For instance, in one particular case, the tidal amplitude was altered deliberately at a midpoint in the dataset. This alteration was made to realistically simulate either a spring tide or a storm surge phenomenon. Such types of uncommon but realistic alterations are employed to make the limitations inherent in the models apparent and real when subjected to abnormal or unexpected conditions. Additionally, such findings are invaluable as they aid in the development of fallback plans or safety nets. In the scenario of production systems, the application of such procedures proves to be absolutely necessary. These provide the continuity of service and functionality even in situations where data distributions experience sudden and unexpected changes.

With respect to the preprocessing step, one specific thing to note is that the level of data cleaning needed was minimized to an absolute minimum mainly because the data involved here were synthetic in nature. Nonetheless, the data processing pipeline was designed with strategy in mind such that it could handle well and accommodate a range of issues such as missing values, outliers, and jitter – these are elements that would undoubtedly become of increasing significance as we proceed toward the use of real sensor data. Additionally, the modularity of this pipeline makes it easy to add various important elements such as filters, interpolators, and imputers, which are essential when operating with raw marine current logs from advanced instruments such as ADCPs. This careful design consideration not only brings capability to the pipeline but also allows for the same feature engineering considerations that have been used here to be easily carried over to real-world applications with no more than minor modifications or adaptations along the way. In addition, it is also important to highlight the fact that the importance of interpretability in the feature engineering process cannot be overstated and is absolutely paramount. In the case of smart buildings, energy choices tend to require human oversight or a detailed audit trail for accountability. A forecasting model that can justify its thought process – for example, by showing that an unexpected forecast tidal current spike is due to the phenomenon of a Full Moon and past observed acceleration trends – will have a much greater level of confidence and trust from facility managers as well as policymakers. This is one of the key reasons why XGBoost, with built-in features such as feature importance scoring in addition to SHAP (SHapley Additive exPlanations) interpretability features, is so useful and relevant in this specific domain. Finally, the outstanding disaggregation and natural modularity of the data simulation and feature engineering stack offer an incredibly valuable template that is readily repurposed for application in other fields of study beyond tidal energy in isolation. In other words, the same new methodologies can be readily transferred to other applications, ranging from predicting roof turbine wind speeds, forecasting solar irradiance for the integration of photovoltaics, to forecasting ambient demand in disaggregated microgrids. Across each of these varied uses, the ability to simulate particular physical processes of interest to the associated field, add variability to the models, and engineer features that are interpretable constitutes a central pillar of intelligent energy planning paradigms. As such, even as the present research is focused primarily upon the investigation of coastal tidal flows, it must be noted that the bottom-up methodology adopted herein is not merely highly generalizable but also scalable to the wider and continually changing smart energy context.

1.6 DESIGN OF FORECASTING MODELS

1.6.1 XGBoost Model

The XGBoost model utilized in this research is an efficient ensemble-based learning algorithm that is particularly suited for structured tabular data. It builds an additive model by aggregating predictions from many weak decision tree learners and then training each subsequent tree to reduce the errors made by the previous tree. Through this methodology, XGBoost is able to learn nonlinear patterns between input features and the target with great predictive accuracy. XGBoost has one of its strongest points in that it is highly computationally efficient, achieving rapid training and low memory usage – making it very appealing for deployment in edge computing settings such as smart buoys or coastal embedded systems.

For tidal prediction, XGBoost input data were created as flat feature vectors extracted from raw time series. All samples contained lagged tidal current speed (velocity_t–1, velocity_t–2, velocity_t–3), cyclically encoded hour-of-day features (Hour_sin, Hour_cos), and a computed power output value as a function of the cube of velocity. This engineered representation enables the model to learn temporal dynamics implicitly without the need for recurrent architecture.

Training was implemented using 80/20 data split, where hyperparameters like learning rate, maximum depth, and number of estimators were tuned using grid search with k-fold cross-validation. The interpretability of the model was improved by using SHAP, which provide a measure of the contribution of each feature to each prediction. Such transparency is useful in actual energy systems, where human operators may be required to authenticate or audit the outputs of the model.

The last XGBoost model performed very well with a root mean squared error (RMSE) of 0.0070 m/s and an R^2 metric of 0.9998 on the test set. Notably, the model file after serialization was lightweight (70 KB), and inference times were always less than 50 ms – suitable for low-latency control loops in smart energy infrastructure.

1.6.2 LSTM Model

The LSTM network provides a radically new strategy for tidal prediction, with its ability to model temporal dependencies directly from raw sequences. As a type of specialized RNN, LSTM is specifically designed to keep track of long-term dependencies and manage non-stationary patterns in time-series data via memory cells and gate mechanisms. This design allows it to learn from sequences of previous observations without having explicit lag features like in conventional models.

In this work, the LSTM was set to input data with a three-dimensional tensor shape of (samples, time_steps, features), where each input sequence had 12 time steps and the same five engineered features as XGBoost. The model structure consisted of stacked LSTM layers followed by dense output layers. The training utilized the Adam optimizer and mean squared error loss function, with early stopping used to prevent overfitting.

Although the LSTM model showed healthy learning behavior during training, its performance on the test set was suboptimal, with an RMSE of 0.2114 m/s and an R^2 score of 0.8121. The LSTM model also presented challenges for deployment. Model size after training (~1 MB) was much greater compared to XGBoost, and average inference time was over 400 ms. These measures indicate that the LSTM is less feasible for deployment in real time or edge-limited environments without significant optimization.

Additionally, LSTM models are also less interpretable compared to decision tree models. In contrast to XGBoost, which provides SHAP values to interpret predictions, LSTM is a black box, and therefore tracing the contribution of individual features to output values is more challenging – an essential limitation in environments where model clarity is a requirement.

All these issues notwithstanding, LSTM is still useful in situations where modeling long-range dependencies and temporal continuity is critical, and there is a sufficient computational budget available. In the future, one might be able to use optimized or compressed forms of LSTM (e.g., quantized model or gated recurrent units (GRU)) to enhance deployment practicability at the cost of sequence modeling capability.

1.7 EXPERIMENTAL SETUP AND EVALUATION METRICS

To test the practical applicability of predicting tidal current speeds for smart coastal structures in the real world, a strong experimental setup was established that balances both statistical robustness and deployment realism. This two-fold evaluation approach guarantees that the selected ML models not only generate effective predictions on the past data but also operate effectively under the physical limitations of embedded systems like microcontrollers, edge artificial intelligence (AI) chips, or low-power marine instruments. In this way, the experiments mimic both a data science validation platform as well as a deployment-focused engineering test bench.

Synthetic data used for experimentation comprised 576 sequential data points, each of which corresponded to a 5-minute sampling interval over a total duration of 48 hours. This time resolution mirrors actual building control use cases, where energy storage discharge, HVAC scheduling, and load shifting can take place every 5–15 minutes. The dataset contains engineered features like lagged velocities, cyclic encodings of time, and computed power outputs, so it is applicable to both flat and sequential models.

A time-based 80:20 data division was utilized, devoting the initial 460 samples to training and leaving the rest of 116 as a test set. This guarantees that model validation takes place on temporally unseen data, replicating future environments and preserving causality. In contrast to random shuffling commonly applied in classification problems, time-series models have to ensure sequence integrity because time-series models have to represent deployment scenarios correctly. This is particularly important for LSTM networks that learn from ordered sequences of inputs.

Hyperparameter tuning was done independently for XGBoost and LSTM. For XGBoost, a grid search technique was used to search through a set of values for learning rate (η), number of estimators (trees), maximum depth per tree, subsample ratio, and regularization terms (λ and α). The optimization criterion was minimizing the RMSE without allowing overfitting via cross-validation. Five-fold cross-validation was used on the training data for generalizability across time segments.

For the LSTM model, the tuning was more subtle because of the computational cost associated with training deep neural networks. Multiple architectural options were experimented with varying amounts of LSTM layers, units per layer (32–128), dropout probabilities, and sequence window sizes. The last model employed a two-layer LSTM structure followed by a dense output layer, which was trained with the Adam optimizer and learning rate of 0.001 and mean squared error loss function. A batch size of 16 was an appropriate balance between stability in learning and memory usage. Early stopping with patience of ten epochs was used, tracking validation loss to avoid overtraining and decrease computation.

The models were validated on two critical aspects: predictive accuracy and deployment effectiveness. Predictive accuracy examines the degree to which a model's output matches up with real values and was measured in terms of two popular statistical metrics: RMSE and the R^2 score (coefficient of determination). RMSE measures the average error of prediction and puts more penalty on large errors, therefore putting more importance on consistency and accuracy. It is mathematically expressed as follows:

$$\text{RMSE} = \sqrt{\frac{1}{n}\sum_{i=1}^{n}\left(y_i - \widehat{y_i}\right)^2} \tag{1.1}$$

Lower RMSE indicates a more robust and precise forecasting model. The R^2 score, however, is the percentage of variance in the target variable (tidal current speed) explained by the independent input variables. It is calculated as follows:

$$R^2 = 1 - \frac{\sum_{i=1}^{n}\left(Y_i - \widehat{Y_i}\right)^2}{\sum_{i=1}^{n}\left(Y_i - \overline{Y}\right)^2} \tag{1.2}$$

A value approaching 1 indicates stronger explanatory capability and improved alignment with trends. For fairness and reproducibility, both metrics were calculated based on a common test dataset for both the XGBoost and LSTM pipelines, establishing an identical benchmark for comparison. Aside from statistical precision, deployment efficiency was gauged through three real-world indicators: model file size, inference time, and memory consumption. Model size, in terms of kilobytes or megabytes, determines the ease of storage and loading onto embedded systems. Inference time, in milliseconds, indicates the speed at which the model can generate predictions – a key factor for real-time control within smart energy systems. Memory use was qualitatively ranked as either low or high to indicate runtime RAM use and hardware loading. Together, these criteria offer a thorough evaluation of both theoretical performance and practical suitability of the forecasting models.

As can be seen from the output, the XGBoost model produced essentially perfect prediction accuracy with an R^2 value of 0.9998, meaning it was able to explain more than 99% of the variation in tidal current speeds. Its RMSE of 0.0070 m/s is a reflection of little deviation from real values and thus is very accurate for short-term prediction purposes. On the other hand, although the LSTM model had picked up some of the underlying sequence patterns, it was behind in accuracy with an RMSE of 0.2114 and a lower R^2 value of 0.8121. The discrepancy could be due to the higher complexity of the LSTM and its need for larger training data to generalize well.

In terms of deployment, XGBoost was much more efficient. Its serialized model size was only 70 KB, low enough to be deployed on microcontrollers like Raspberry Pi Pico, ESP32, or low-power ARM Cortex processors. Further, its inference latency was always less than 50 ms, which is within the control loop window for most BEMS. Memory consumption was negligible, making it appropriate even for battery-operated or solar-powered edge nodes.

LSTM, however, was problematic for lightweight deployment. Its model size was over 1 MB, and inference times were commonly over 400 ms. Such performance renders it hard to utilize in real-time applications except in conjunction with powerful computing resources or in the presence of hard model compression strategies. In addition, LSTM does not have built-in explainability, making it less appealing for regulatory purposes and human-in-the-loop decision-making scenarios.

These findings highlight the importance of selecting forecasting models that trade-off predictive performance and deployability. While the LSTM model can tap into long-range dependencies due to its sequential nature, it comes with the cost of increased computational complexity and memory requirements – considerations that impact its suitability for edge-deployment scenarios.

On the other hand, XGBoost presented an attractive combination of rapid inference time, low resource consumption, and high predictive performance. This makes it most appropriate for real-time forecasting in embedded controllers used in coastal smart building applications. The relative comparison of the two models is summarized in Table 1.1.

TABLE 1.1 Comparison of XGBoost and LSTM models for forecast accuracy and deployment efficiency in coastal energy systems

METRIC	*XGBOOST*	*LSTM*
RMSE	0.0070	0.2114
R^2 score	0.9998	0.8121
Model file size	70 KB	~1 MB
Inference time	<50 ms	>400 ms
Memory usage	Low	High

As shown in Table 1.1, XGBoost had an R^2 value of 0.9998 and an RMSE of 0.0070 m/s, indicating near-perfect match with actual tidal current data. It also took less than 50 ms for inference and consumed just 70 KB of memory. As compared to this, the LSTM model had an RMSE of 0.2114 and an R^2 value of 0.8121, which indicates comparatively lower precision. Its higher memory consumption (~1 MB) and slower inference time (>400 ms) also reduce its suitability to low-power, real-time applications. These findings confirm the significance of model transparency and efficiency in live contexts, particularly for building energy systems installed at the coastal edge.

On a deployment front, XGBoost was much more efficient. Its serialized model size was only 70 KB – small enough for deployment on microcontrollers like Raspberry Pi Pico, ESP32, or low-power ARM Cortex processors. Additionally, its inference latency was always below 50 ms, which is comfortably within the control loop window for the majority of BEMS. Memory usage was minimal, making it viable even for battery-powered or solar-backed edge nodes.

LSTM, however, was challenging for lightweight deployment. Its model size was well over 1 MB, and inference took longer than 400 ms in most cases. Such performance is not feasible to use in time-critical situations unless complemented with high-end computing resources or compressed aggressively using model compression methods. Additionally, LSTM does not have explainability built into the algorithm, making it less suitable for regulatory and human-in-the-loop decision-making applications.

These observations strengthen the necessity for model selection to reconcile predictive complexity with engineering practicability. Although LSTM has potential in situations where deep temporal modeling is critical – e.g., long-horizon prediction or smooth signal following – its resource requirements are presently beyond its advantages for embedded deployment. XGBoost, on the other hand, presents itself as a deployable and feasible solution with high accuracy, low resource usage, and model explainability – prime demands for smart building energy systems deploying at the coastal edge.

1.8 RESULTS AND DISCUSSION: WHAT WORKS BEST FOR SMART BUILDINGS?

1.8.1 Accuracy Insights

The relative performance of XGBoost and LSTM exhibits stark contrasts in terms of how well each of them captures the dynamics of tidal current fluctuation, which is key to powering smart building applications in a reliable manner.

XGBoost performed extremely well on the test set with an RMSE of only 0.0070 m/s and an R^2 score of 0.9998. The scores are proof that the model was exceptionally good in learning and predicting the cyclic patterns in tidal flows with engineered features. Most importantly, it was able to trace the rising and falling patterns of the currents with almost zero error.

Conversely, the LSTM model, though designed for sequential data, returned an RMSE of 0.2114 m/s and an R^2 measure of 0.8121. While LSTM was able to capture broad trends, it fared poorly in following sharp or sudden changes in tidal speed. This weakness may stem from its increased sensitivity to noise or from the small size of the training dataset employed in modeling. Visual comparisons support this determination: Figure 1.2 illustrates the way XGBoost's predicted curves correspond highly with actual observed values, even near the peak and trough points. Figure 1.3 illustrates the smoother predictions of the LSTM, commonly lagging behind during high-speed-changing tide events.

Such predictions are particularly vital for coastal smart buildings, which can be energy-scheduled based on rapid and precise short-term forecasts, such as between tidal power, storage, or grid usage.

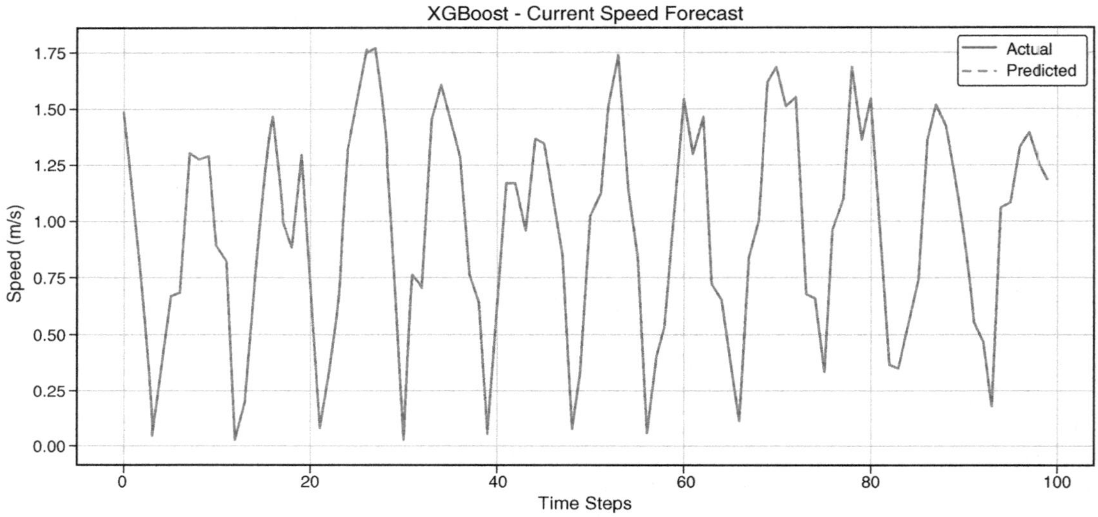

FIGURE 1.2 XGBoost forecast accuracy: alignment of predicted and actual tidal energy curves.

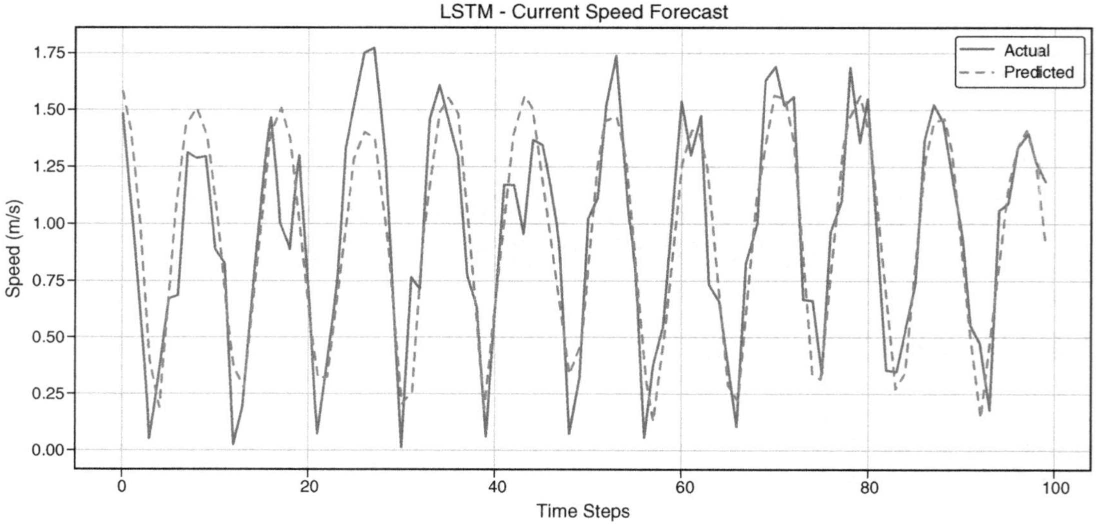

FIGURE 1.3 LSTM forecast behavior: smooth predictions with temporal lag during rapid tidal variations.

For these time-critical applications, prediction lag or over-smoothed predictions may result in less-than-optimal control decisions, decreasing system efficiency.

Whereas LSTM might be more appropriate for situations with extended prediction time horizons and availability of larger data sizes, XGBoost is the stronger and more stable model for real-time, edge-centric applications common in intelligent building management systems.

1.8.2 Deployment Readiness

A model's precision is only half the equation with regard to smart building integration. So, the model performs only half as much as how well that model will actually perform when deployed in real-world environments. In edge-based architectures such as in coastal smart buildings, memory, processing power, and response-time constraints are paramount.

XGBoost exhibited outstanding deployability traits. With its model size in serialized form around 70 KB and very low CPU overhead, it consumes negligible memory and storage. Inference latency for XGBoost was below 50 ms on baseline hardware, making it real-time responsive to energy management. Its small-footprint architecture and support for low-power processors further establish its feasibility in embedded devices typically utilized in building control systems.

On the other hand, although more complex in its architecture, the LSTM model exhibited significant deployment issues. The model, kept in .h5 format, occupied a file size of approximately 1 MB. Additionally, it required much more memory and computational power for both training and inference. The average inference time was in excess of 400 ms, which, though fine in offline environments, might introduce considerable delays in real-time control applications in the absence of GPU acceleration.

This balance between model complexity and deployment efficiency is core to real-world decision-making. The capacity of LSTM to capture long-term dependencies is at the expense of increased model sizes and reduced responsiveness. For constrained hardware in smart buildings that require immediate forecast-based decisions, such latency and overhead may present operational bottlenecks.

Therefore, XGBoost not only presents itself as a more precise model in this experiment, but also as a more feasible choice for deployment in the real world – particularly where low complexity, high speed, and minimal memory are critical. LSTM, as important as it is in high-resource environments or for batch forecasting applications, needs further tuning toward edge deployment in the context of smart building energy systems.

1.9 DEPLOYMENT BLUEPRINT FOR SMART BUILDINGS

Successful integration of tidal forecasting in smart coastal buildings is not merely about having accurate predictions but rather about a smooth deployment path that integrates prediction outputs into real-time energy management systems. The following describes a workable roadmap for incorporating forecasting models into BEMS.

The key concept is to make the tidal forecasting model a modular element in the smart building's control system. Forecasting executes on a local edge device (e.g., Jetson Nano or Raspberry Pi), with real-time sensor data from tide-measuring sensors or cloud-based APIs. The model acts on this data, producing predictions of tidal current speed and accompanying power generation potential.

These predictions are then sent to the BEMS, which translates the output to make control decisions such as whether to charge a battery, use the grid, or optimize appliance load. It also has to take into account other inputs such as occupancy, weather prediction, or electricity prices.

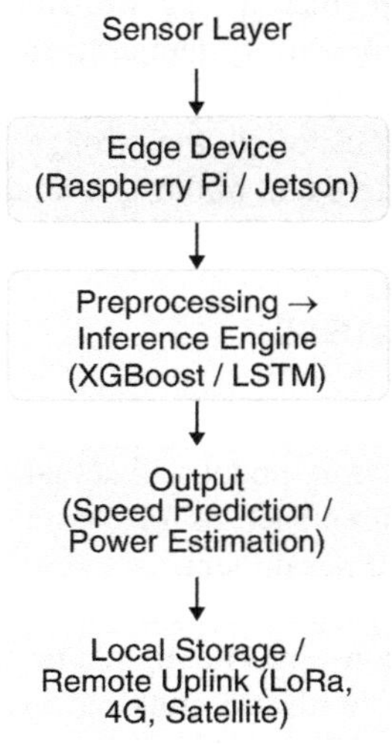

FIGURE 1.4 Comprehensive system pipeline for smart coastal building energy management.

To facilitate this process:

Real-time tidal measurements are received by data acquisition modules.
Preprocessing units normalize and format the input.
The ML model (XGBoost or LSTM) produces the prediction.
Forecast outputs are combined into a decision engine in connection with actuators and controllers within the building.
This end-to-end system pipeline, from raw data ingestion to energy decision-making within the smart building ecosystem, is shown in Figure 1.4.

It is important to note that major characteristics of this architecture are modularity, low-latency communication, and re-trainability of the model. Additionally, with the model run locally, data privacy is maintained, and response times are reduced, which is important for time-sensitive control decisions.

This deployment pipeline not only drives energy forecasting with the power of ML but actually makes intelligent, real-time renewable energy management for coastal buildings.

1.10 CHALLENGES AND DESIGN CONSIDERATIONS

A tidal forecasting system based on ML in coastal smart buildings has potential advantages, but there are a number of real-world practical challenges that need to be overcome for sustained success and operational reliability.

1.10.1 Hardware Limitations

Smart buildings tend to use edge devices with inferior computational capabilities like single-board computers (e.g., Raspberry Pi) or microcontrollers. These platforms are limited by processing power, memory, and power consumption. While XGBoost can run effectively on such hardware, the deployment of deep learning models such as LSTM usually demands more advanced processors or specialized GPUs that might not be available in all situations. Coping with heat dissipation and supporting round-the-clock uptime in exposed coastal or weather-exposed remote environments only adds to these hardware-specific difficulties.

1.10.2 Compatibility with Building Automation Protocols

For real-time control of energy consumption, the forecasting system needs to be integrated with current BEMS. The systems use protocols like BACnet, Modbus, or KNX. Compatibility will require model outputs being converted into common control signals that can be used to activate devices. Technical and domain knowledge in smart building systems and automation standards is needed to build such interfaces.

1.10.3 Model Retraining and Drift Handling

Coastal environmental conditions are dynamic, which are subjected to seasonal variability, climate fluctuations, and even structural changes to adjacent water bodies. In time, this can result in data drift, where performance of the model suffers from changes in underlying patterns. Repeated retraining of models is necessary to ensure accuracy. Using a feedback loop by which the system can track performance deviations and trigger retraining (automatically or by hand) is an important factor in future-proofing deployment. Resource-frugal retraining techniques will also be important, particularly where models are being updated offline without the support of the cloud.

Meeting these challenges is not a technical exercise but an integrated design consideration that calls for coordination of ML engineers, building system integrators, and energy managers to facilitate the seamless adoption of AI-based energy forecasting in real-world smart infrastructure.

1.11 CONCLUSION AND FUTURE OUTLOOK

This chapter investigated the use of ML, namely XGBoost and LSTM, to predict tidal energy output to assist smart energy planning of coastal structures. By thorough experimentation and performance comparison, the most pragmatic solution proved to be XGBoost. It acts as a strong combination of prediction accuracy, minimum memory usage, and extremely fast inference times. These qualities render it most suitable for real-time operation on resource-limited edge devices common in smart building infrastructure.

LSTM, while it can represent intricate sequential dependencies in tidal flow data, showed considerable overhead in terms of computational requirement and inference delay. While it could be more appropriate for situations calling for larger forecast windows or finer learning granularity, its deployability in real-time applications is still in question without considerable hardware assistance or optimization.

In the future, there are some attractive research directions that can make such forecasting systems more robust and adaptable. Hybrid frameworks blending the explainability of tree methods with deep learning's temporal modeling capabilities may provide improved performance trade-offs. Federated learning frameworks might also allow for the simultaneous collective training of forecasting models across various coastal buildings without compromising data privacy. Lastly, the integration of automated model drift detection and retraining pipelines will be critical to adjust to evolving environmental conditions over time.

Finally, plugging forecasting intelligence into the control loop of energy-efficient buildings is a leap forward in sustainable coastal infrastructure, where ML doesn't only predict the tides but assists buildings to react to them smartly.

BIBLIOGRAPHY

Aljber, M., Jeong, J.S., Cabrera, J.S., Calvo, M.A.S., Chisale, S.W., Williams, Z. and Lee, H.S., 2024. Optimal site selection and potential power assessment for tidal power generation in the Seto Inland Sea, Japan, based on high-resolution ocean modelling and multicriteria analysis. *Applied Energy*, 372, 123843.

Adcock, T.A., Draper, S., Willden, R.H. and Vogel, C.R., 2021. The fluid mechanics of tidal stream energy conversion. *Annual Review of Fluid Mechanics*, 53(1), 287–310.

Ali, A.F.D.H., Rosli, R. and Basunia, M.A., 2023. Tidal energy in Brunei Darussalam: Motivations, potentials and challenges. *AIP Conference Proceedings*, 2643(1), 050073.

Alvarez, E.A., Manso, A.N., Gutiérrez-Trashorras, A.J., Francos, J.F. and Secades, M.R., 2013. Obtaining renewable energy from tidal currents in the Aviles port: New services for citizens. In: *2013 International Conference on New Concepts in Smart Cities (SmartMILE)*, Gijón, Spain. IEEE.

Auguste, C., Nader, J.R., Marsh, P. and Cossu, R., 2019. Influence of tidal energy converters on sediment dynamics in tidal channel. In: *Proceedings of 13th European Wave and Tidal Energy Conference*, Naples, Italy, pp. 1–6.

Badwawi, R.A., Abusara, M. and Mallick, T. 2015. A review of hybrid solar PV and wind energy system. *Smart Science*, 3(3), 127–138.

Bricker, J.D., Esteban, M., Takagi, H. and Roeber, V., 2017. Economic feasibility of tidal stream and wave power in post-Fukushima Japan. *Renewable Energy*, 114, 32–45.

Buhali, M.L., Abundo, M.L.S. and Ang, M.R.C.O., 2012. Site selection procedures for tidal in-stream energy in the Philippines: A preliminary study. In: *2012 10th International Power & Energy Conference (IPEC),* Ho Chi Minh City, Vietnam, pp. 110–114. IEEE.

Chatzirodou, A., Karunarathna, H., Mainland, S., Firth, P. and Park, S., 2014. Impacts of tidal energy extraction on sea bed morphology. In: *Proceedings of 34th Conference on Coastal Engineering*, Seoul, Korea.

Chen, H.H. and Lee, A.H., 2014. Comprehensive overview of renewable energy development in Taiwan. *Renewable and Sustainable Energy Reviews*, 37, 215–228.

Chozas, J.F. and Soerensen, H.C., 2009. State of the art of wave energy in Spain. In: *2009 IEEE Electrical Power & Energy Conference (EPEC)*, Montreal, Quebec, Canada. IEEE.

Coiro, D.P., Troise, G. and Bizzarrini, N., 2018. Experiences in developing tidal current and wave energy devices for Mediterranean Sea. *Frontiers in Energy Research*, 6, 136.

Coles, D., Angeloudis, A., Greaves, D., Hastie, G., Lewis, M., Mackie, L., McNaughton, J., Miles, J., Neill, S., Piggott, M. and Risch, D., 2021. A review of the UK and British Channel Islands practical tidal stream energy resource. *Proceedings of the Royal Society A*, 477(2255), 20210469.

Cradden, L., Kalogeri, C., Barrios, I.M., Galanis, G., Ingram, D. and Kallos, G., 2016. Multi-criteria site selection for offshore renewable energy platforms. *Renewable Energy*, *87*, 791–806.

Cui, Y., Xu, D., Liu, Z. and Ross, D., 2020. Global research and trends in renewable energy: Ocean waves, tidal energy and offshore wind. *Journal of Coastal Research*, 95(SI), 1485–1489.

Dalton, G., Allan, G., Beaumont, N., Georgakaki, A., Hacking, N., Hooper, T., Kerr, S., O'Hagan, A.M., Reilly, K., Ricci, P. and Sheng, W., 2015. Economic and socio-economic assessment methods for ocean renewable energy. *Renewable and Sustainable Energy Reviews*, 45, 850–878.

Dawn, S., Gope, S., Das, A., Bhowmik, D. and Koley, I., 2019. Tidal energy as emergent energy source: A review. *International Journal of Computational Intelligence & IoT*, 2(1), 340–345.

Serverless File Management System Using AWS S3 and Lambda for Scalable and Secure Storage

2

Kollipara Anirudh, J. Muthu Bharath,
Tharaga Sharmilan, and O. V. Gnana Swathika

2.1 INTRODUCTION

Static website hosting has gained traction as a cost-efficient and low-maintenance alternative to dynamic web hosting. Unlike their dynamic counterparts, static websites deliver content stored in its original form without server-side processing, making them suitable for applications with relatively stable content requirements. Amazon Simple Storage Service (S3) offers a reliable and scalable platform for static site hosting while providing integration with a wide range of Amazon Web Services (AWS).

This study explores the implementation of a secure static website hosted on AWS S3. Key objectives include the development of client-side authentication mechanisms, secure file upload functionality, and the maintenance of cost-efficiency within AWS Free Tier limitations.

DOI: 10.1201/9781003649700-2

2.2 MOTIVATION

The need for cost-efficient and secure cloud solutions has driven the adoption of serverless architectures. While S3 static website hosting addresses scalability and reliability, integrating security measures such as client-side authentication presents challenges that must be addressed for real-world deployments.

2.2.1 Background

2.2.1.1 Static website hosting

Static websites are characterized by their straightforward architecture, as they serve pre-rendered files directly to users without requiring server-side execution. This simplicity reduces overhead, enhances performance, and minimizes operational costs. Key features of static websites include:

1. **Direct Content Delivery**: Files are served as stored.
2. **Reduced Complexity**: Absence of server-side logic simplifies architecture.
3. **Cost-Effectiveness**: No server runtime costs are incurred.
4. **Use Cases**: Suitable for blogs, documentation, and file-sharing portals.

2.2.1.2 AWS S3 architecture

Amazon S3 is a globally distributed object storage service supporting use cases such as static site hosting, file backups, and scalable storage solutions. Key architectural components for static hosting include:

1. **Bucket Configuration**: Hosting is enabled at the bucket level, with options to set index and error documents.
2. **Endpoint Support**: Region-specific website endpoints differ from REST API endpoints for object access.
3. **Access Control**: Bucket policies, Cross-Origin Resource Sharing (CORS) configurations, and Identity and Access Management (IAM) roles ensure secure operations.
4. **Custom Domains**: S3 integrates with services like AWS Route 53 to support branded URLs.

2.2.1.3 Detailed technical architecture

The technical architecture for secure static website hosting on AWS S3 incorporates various AWS services to provide a secure, scalable, and low-cost solution for file hosting. The architecture begins with an Amazon S3 bucket, configured specifically for static website hosting, where files are served directly to end-users without server-side rendering. This setup enables efficient handling of stable, pre-rendered content. Bucket policies are carefully crafted to allow public read access only, ensuring sensitive operations like uploads are restricted to authenticated users. Using AWS IAM, specific roles are assigned to control permissions—enabling secure, fine-grained access to resources, such as generating pre-signed URLs for upload. These pre-signed URLs allow users to securely upload files up to 4 MB without needing a direct connection to backend servers, reducing complexity and server costs. JSON Web Tokens (JWT) are used for client-side authentication, where a signed token validates the user's identity before permitting access to certain operations. To monitor activity and ensure compliance with security policies, AWS CloudTrail and CloudWatch are configured to log access and API calls, enabling a robust monitoring system. Finally, lifecycle policies are applied to manage object storage cost by transitioning older data to lower-cost storage options like S3 Glacier, maintaining overall cost-effectiveness.

2.3 IMPLEMENTATION

2.3.1 Design Goals

1. Enable secure, scalable static site hosting using AWS S3:
   - Ensure the S3 bucket is configured for high availability and fast delivery of static assets.
   - Optimize settings to minimize costs while staying within the AWS Free Tier constraints.
2. Provide a client-side authentication mechanism:
   - Authenticate users before granting access to sensitive operations such as file uploads.
   - Use JWTs to minimize server overhead and maintain stateless operations.
3. Facilitate secure file uploads for authenticated users:
   - Leverage pre-signed URLs to delegate file upload permissions securely

2.3.2 Advanced Testing and Monitoring

To ensure reliable performance and security of the AWS S3 static hosting setup, advanced testing and monitoring strategies are implemented. Testing involves evaluating the application under different network conditions, file sizes, and loads to gauge responsiveness and stability. Load testing is particularly crucial for identifying performance bottlenecks when handling multiple concurrent uploads. Tools like Apache JMeter and AWS CloudWatch Synthetics can simulate user traffic, testing upload limits, page load times, and error handling for larger datasets and increased activity. CORS and security configuration tests are also performed to verify that permissions are correctly restricted and data exposure risks are minimized.

On the monitoring side, AWS offers a range of tools to track application health, usage, and compliance. AWS CloudWatch provides real-time insights into the infrastructure's performance by monitoring metrics like bucket request rates, error rates, and latency. Custom CloudWatch metrics are created to track specific actions like file upload progress, average file size, and authentication times. CloudWatch Alarms are set up to notify administrators if thresholds are crossed, such as upload failures exceeding a set percentage, which could indicate security issues or configuration problems.

AWS CloudTrail is utilized to log API-level activity within the S3 bucket, tracking requests like GET, PUT, and LIST operations. This helps audit user actions and ensures compliance with security policies by monitoring unauthorized access attempts. To further enhance security, AWS Config can be used to monitor S3 configurations continuously, ensuring policies stay aligned with best practices for restricted access and encryption.

Finally, automated alerts via Amazon SNS (Simple Notification Service) provide immediate notifications in case of critical events, such as high error rates, unusual API activity, or cost anomalies. These alerts are sent via SMS or email to designated administrators, allowing quick responses to potential issues. Dashboard visualization tools in CloudWatch help teams analyze trends and maintain an overview of performance, making it easier to proactively address issues and optimize resource allocation.

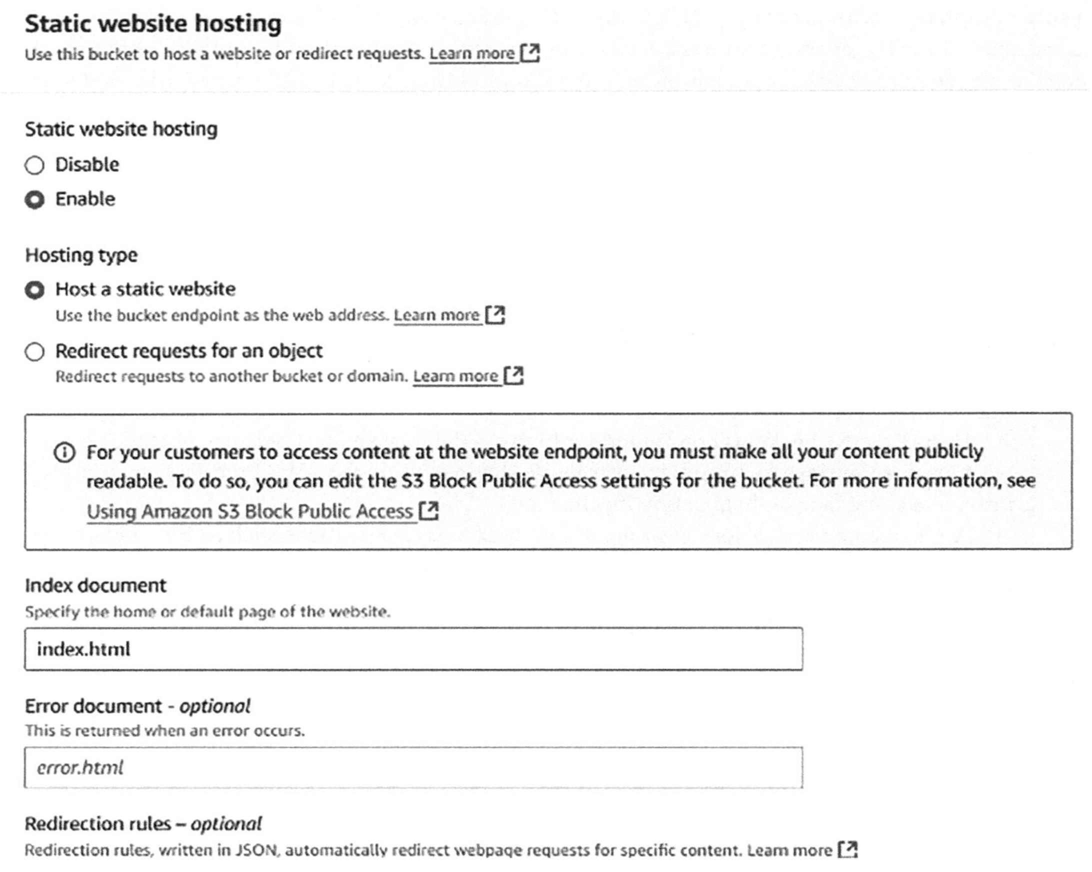

FIGURE 2.1 AWS S3 console—static website hosting configuration panel illustrates the web interface for configuring S3 static website hosting, highlighting key settings such as enabling hosting, specifying the index document (index.html) as the default page, error document for custom error pages, and the note about requiring public read access for website functionality.

This testing and monitoring framework collectively ensures that the static website hosting setup is secure, cost-effective, and resilient under varying conditions as shown in Figure 2.1.

2.3.3 Comparative Analysis

When choosing a cloud storage solution for static website hosting, AWS S3 stands out for its robust scalability, high availability, and integration with other AWS services. However, alternatives like Google Cloud Storage and Microsoft Azure Blob Storage also offer competitive features, making it essential to analyze their differences. AWS S3 excels in seamless integration with the broader AWS ecosystem, allowing for efficient integration with services like CloudFront for content delivery, Lambda for serverless processing, and IAM for granular access management. AWS S3 also has a comprehensive Free Tier for low-cost implementations, along with fine-tuned cost controls like lifecycle policies for automatic data transitions to cheaper storage classes (e.g., Glacier).

On the other hand, Google Cloud Storage offers strong multi-regional redundancy by default, allowing data replication across regions to improve reliability without additional configuration. Google also provides a user-friendly setup for CORS configuration, which simplifies setting permissions for public and cross-origin access. Google Cloud's Nearline and Coldline storage classes parallel AWS's Glacier options, though they are sometimes more cost-effective for infrequent access needs.

Microsoft Azure Blob Storage is particularly notable for its integration with enterprise tools, such as Active Directory, making it ideal for large organizations with existing Microsoft infrastructure. Azure's storage options—Hot, Cool, and Archive tiers—cater to various access needs similarly to AWS but also include granular pricing models based on data retrieval speeds. Additionally, Azure provides Blob Indexer, a unique feature that can add searchable metadata to stored objects, useful for applications requiring advanced search capabilities.

In terms of pricing, AWS offers the most flexible pricing structure through its Free Tier, but Google and Azure are competitive, with occasional advantages in certain regions and for specific storage tiers. In summary, AWS S3 is generally preferred for projects that need extensive scalability and integration with AWS services, while Google Cloud Storage is known for cross-regional availability, and Azure Blob Storage is optimal for enterprises that prioritize Microsoft ecosystem integration. Each service has distinct strengths, so the choice depends on specific use cases, budget constraints, and ecosystem compatibility.

2.3.4 Use Cases

AWS S3's static website hosting with client-side authentication offers versatile applications across various industries, particularly for scenarios requiring reliable, low-cost hosting with moderate security needs. Educational institutions can leverage this setup for hosting static content such as course materials, tutorials, and reference documents, making them easily accessible for students without extensive backend management. Small businesses can use S3 to host landing pages, product catalogs, or portfolios, benefiting from the cost-effective setup and scalability, allowing them to handle variable traffic with minimal maintenance. Nonprofit organizations can create static informational sites or resource libraries, enabling global access to critical information while keeping costs low through AWS's Free Tier and cost optimization strategies. Additionally, startups can take advantage of S3's flexibility for rapid deployment of static sites during product launches, marketing campaigns, or user onboarding—handling bursts of traffic without requiring complex infrastructure.

For more advanced use cases, S3's ability to integrate with services like AWS Lambda allows Media Companies to deliver multimedia files with minimal latency, providing high-quality experiences to end-users even under high traffic. Similarly, developers and content creators can benefit from S3's support for CORS to host assets like images, videos, or documents that can be embedded across platforms, supporting dynamic linking and distribution.

Each of these cases demonstrates the adaptability of AWS S3 for a wide range of applications that prioritize cost-effectiveness, simplicity, and scalability, especially when dynamic server-side signed URLs make it a secure choice for authenticated file access, ideal for applications requiring user-specific content delivery.

Microsoft Azure Blob Storage is particularly notable for its integration with enterprise tools, such as Active Directory, making it ideal for large organizations with existing Microsoft infrastructure. Azure's storage options—Hot, Cool, and Archive tiers—cater to various access needs similarly to AWS but also include granular pricing models based on data retrieval speeds. Additionally, Azure provides Blob Indexer, a unique feature that can add searchable metadata to stored objects, useful for applications requiring advanced search capabilities.

2.3.5 Security Enhancements

To strengthen security for AWS S3 static website hosting, several key enhancements can be applied to protect both the content and user interactions. First, integrating AWS Cognito provides robust user identity management, enabling multi-factor authentication (MFA) and secure login without developing a custom authentication system. For finer control over access, AWS IAM policies should be configured with the principle of least privilege, granting minimal permissions necessary for users and services to interact with the S3 bucket. By applying bucket policies and Access Control Lists (ACLs), it's possible to ensure that only authenticated users can upload files or access restricted areas, while public resources are isolated.

To further safeguard file uploads as in Figure 2.2, pre-signed URLs allow time-limited access, enabling only authenticated users to perform specific actions like uploading files within a predefined window. This approach prevents unauthorized users from manipulating bucket contents directly. Implementing as in Figure 2.3, CORS policies carefully restrict which origins can access resources, mitigating potential cross-site request forgery (CSRF) attacks.

Data protection is also enhanced with encryption strategies. Enabling S3 bucket encryption ensures that all data at rest is protected, with options for both Server-Side Encryption (SSE) and Client-Side Encryption (CSE), depending on specific compliance needs. Using AWS Key Management Service (KMS) allows for additional control over encryption keys, with automatic rotation for enhanced security. AWS CloudTrail and CloudWatch monitoring provide essential visibility into access patterns, enabling administrators to detect and respond to suspicious activity through real-time alerts. Additionally, enabling AWS Config allows continuous monitoring of bucket security settings, ensuring compliance with security best practices and alerting administrators of any unauthorized changes. Collectively, these security enhancements create a robust and layered defense, ensuring that only authorized users can access sensitive content and minimizing exposure to potential threats.

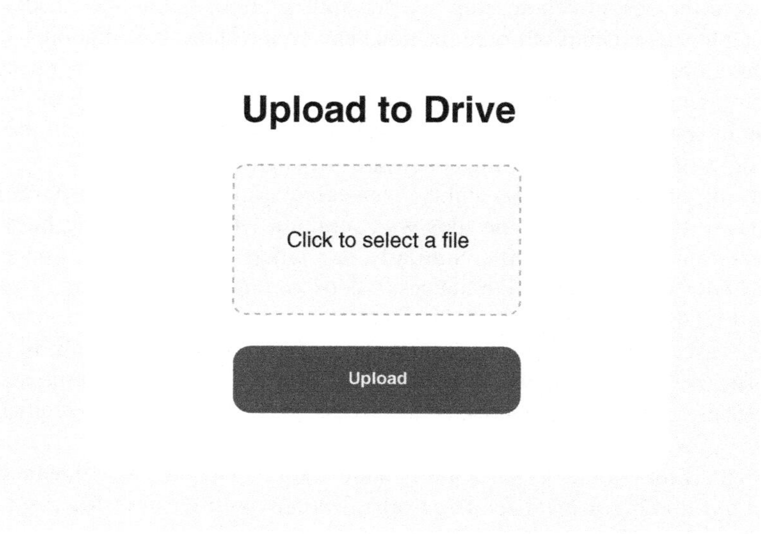

FIGURE 2.2 File upload interface for S3 static website—the screenshot shows the user-facing upload interface with options to either drag-and-drop files or click to select files from the local device, providing a seamless user experience for secure file uploads to AWS S3.

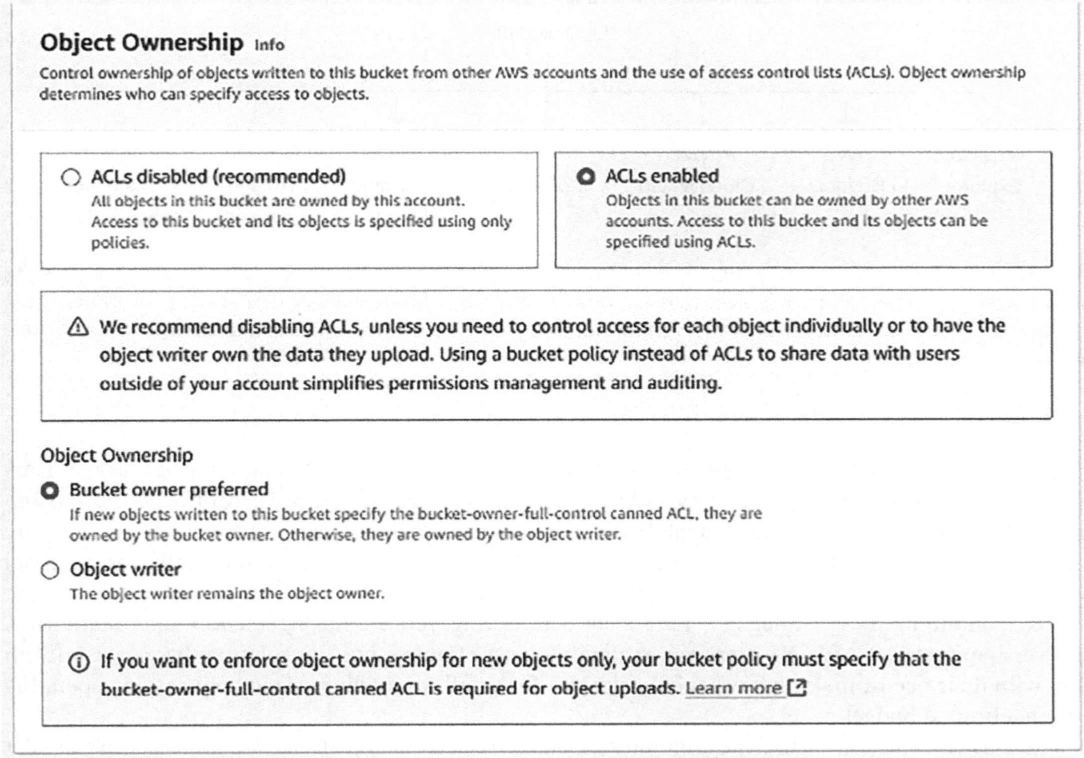

FIGURE 2.3 AWS S3 object ownership configuration—the screenshot shows the S3 console interface for configuring object ownership settings, highlighting the recommended security practice of disabling ACLs and using bucket owner preferred ownership to maintain centralized access control through bucket policies rather than individual object ACLs.

2.3.6 Cost Optimization Strategies

To keep AWS S3 hosting costs low, especially within the Free Tier, several cost optimization strategies as in Figure 2.4 can be implemented to make the most of available resources while avoiding unnecessary expenses. First, Lifecycle Policies should be configured to automatically transition older or infrequently accessed data to cheaper storage classes, such as S3 Glacier or S3 Glacier Deep Archive. This is ideal for archived files or backup data that does not require frequent access, reducing long-term storage costs. Request management is another key strategy; configuring caching for frequently accessed objects using Amazon CloudFront significantly reduces the number of GET requests directly to the S3 bucket, cutting down on request-based charges.

Compression of assets like images, documents, and scripts before uploading to S3 not only minimizes storage costs but also improves loading times, enhancing user experience. Implementing data transfer limits within IAM policies helps control cross-region traffic, preventing unnecessary data egress charges by restricting access to users within specific regions or utilizing only one designated bucket for each region.

Monitoring and Alerts through AWS CloudWatch offer proactive cost management by allowing administrators to track usage metrics in real time. Setting up AWS Budgets and creating alerts for

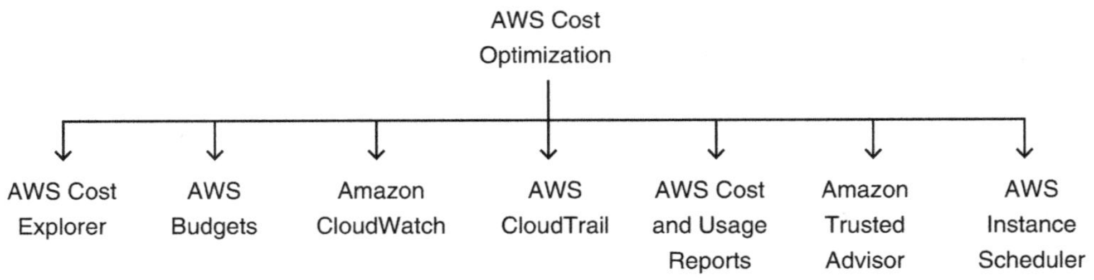

FIGURE 2.4 AWS cost optimization tools ecosystem—the diagram illustrates the comprehensive suite of AWS cost management tools available for monitoring and optimizing cloud spending, including Cost Explorer for analytics, CloudWatch for monitoring, Budgets for alerting, and Trusted Advisor for best practice recommendations.

specific cost thresholds help prevent unexpected charges by sending notifications when usage nears Free Tier or budget limits. Additionally, object versioning should be managed carefully; while versioning provides backup and recovery capabilities, it also increases storage costs if too many versions are retained. Lifecycle policies can automatically delete older versions, ensuring storage efficiency without unnecessary expense.

By combining these strategies—intelligent data tiering, request management, usage monitoring, and version control—AWS S3 users can maintain a cost-effective hosting solution that scales affordably with their needs, making it especially suitable for small- to medium-sized applications operating within a limited budget.

To further refine cost-efficiency, monitoring storage and access patterns can guide more granular adjustments to S3 configurations. For example, periodically reviewing data access frequency helps identify opportunities to move seldom-accessed content to S3 Intelligent-Tiering, which automatically adjusts storage class based on usage, balancing cost with accessibility. Using custom metrics in AWS CloudWatch enables detailed tracking of frequently accessed files, helping identify "hot" objects that can be cached or replicated closer to users with Amazon CloudFront, minimizing direct S3 access and lowering data retrieval charges. Additionally, Data Lifecycle Management can automate the clean-up of temporary or outdated files by deleting them after a specified period, effectively controlling the storage of unnecessary data and keeping monthly costs low.

Optimizing the file upload process also impacts cost; using pre-signed URLs allows authenticated users to upload directly to S3, reducing the need for additional compute services to handle uploads, thus minimizing both resource consumption and operational costs. Scheduled data analysis through AWS Cost Explorer provides insights into long-term storage trends, identifying potential optimizations based on historical usage patterns and suggesting savings plans that could further reduce overall expenses. For organizations with heavier traffic, Reserved Capacity or S3 Storage Class Analysis can predict storage needs and optimize configurations for long-term growth. Combined, these strategies as in Figure 2.5 maximize the cost-effectiveness of AWS S3 without sacrificing performance or accessibility, making S3 an ideal choice for scalable, budget-conscious storage solutions.

2.3.7 Scalability and High-Durability Strategies

AWS S3 is designed for virtually unlimited scalability, supporting applications of any size, from small projects to massive enterprise solutions. To ensure seamless scalability, S3 automatically manages capacity adjustments as in Figure 2.6 based on incoming traffic, allowing it to handle sudden increases in requests without manual intervention. Leveraging Amazon CloudFront, AWS's global

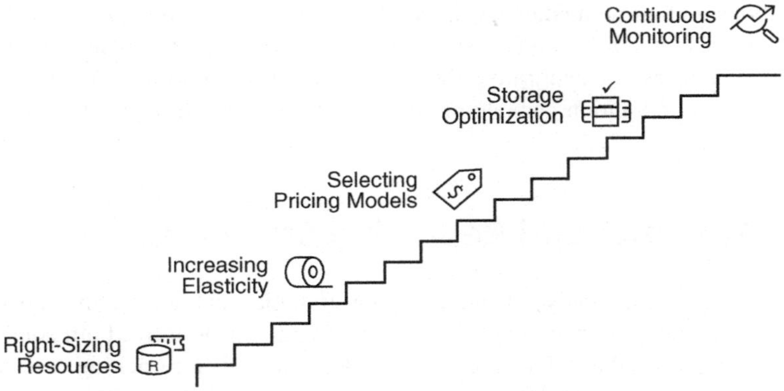

FIGURE 2.5 AWS cost optimization best practices framework—the diagram illustrates the core principles of AWS cost optimization, including right-sizing resources, increasing elasticity through auto-scaling, leveraging appropriate pricing models (such as spot instances or savings plans), optimizing storage classes, and implementing continuous monitoring and improvement cycles.

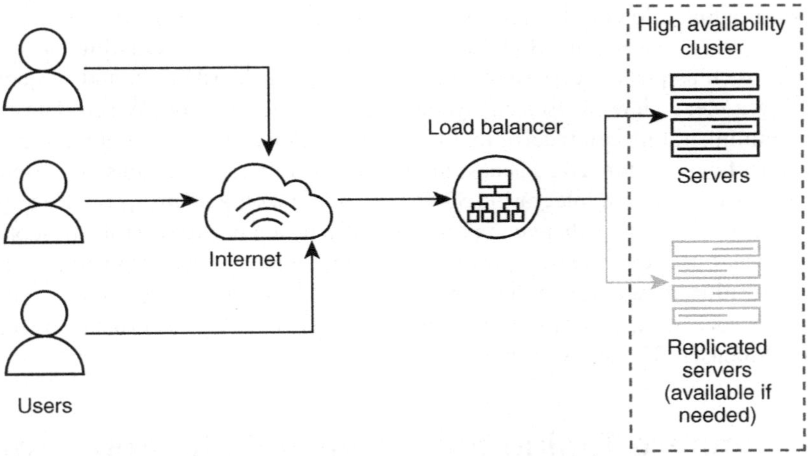

FIGURE 2.6 Load balancer architecture for high availability—the diagram illustrates a load balancing configuration where incoming internet traffic is distributed across multiple servers, with replicated servers available on standby to handle failover and ensure continuous service availability during high load or server failures.

content delivery network (CDN) further enhances scalability by caching frequently accessed files at edge locations worldwide, reducing latency and offloading traffic from the S3 bucket. For high durability, S3 offers 99.999999999% (11 nines) of data durability by storing copies of each object across multiple Availability Zones (AZs) within a region. This multi-zone redundancy ensures that even if one AZ fails, the data remains available.

To protect against regional disruptions and to support disaster recovery, Cross-Region Replication (CRR) can be enabled, creating duplicate copies of data in a secondary AWS region, allowing for faster failover and enhanced disaster resilience. Versioning in S3 is another critical durability strategy, as it enables the storage of multiple versions of objects, preventing accidental deletions and overwrites by retaining a history of file changes. Combined with Object Lock for WORM (Write Once, Read Many) storage, S3 ensures that critical data cannot be deleted or altered before a defined retention period.

For organizations with growing data needs, S3 Intelligent-Tiering optimizes costs without sacrificing durability or accessibility by automatically moving data between frequent and infrequent access tiers based on usage patterns. By integrating these scalability and durability strategies, AWS S3 offers a reliable and future-proof solution capable of supporting high-growth applications with continuous, uninterrupted access to data.

2.3.8 User Analytics and Reporting Strategies

AWS S3 supports powerful user analytics and reporting strategies that provide insights into user behavior, access patterns, and performance trends, helping organizations optimize both content delivery and security. S3 Access Logs and AWS CloudTrail are foundational tools for tracking user interactions with S3 buckets. S3 Access Logs record details of each request made to a bucket, capturing information such as requester IP addresses, request types, and object access frequency. These logs can be stored in a separate S3 bucket for further analysis. CloudTrail, on the other hand, provides a detailed record of API-level actions taken on S3 resources, which is essential for monitoring sensitive operations, auditing user actions, and ensuring compliance with security policies.

To derive actionable insights from raw log data, services like Amazon Athena can be used to query S3 Access Logs directly, allowing for fast, SQL-based analysis without requiring data extraction or additional infrastructure. AWS QuickSight can then visualize this data, creating custom dashboards that track metrics such as peak access times, geographical user distribution, and frequently accessed objects, providing a clear view of user engagement. By setting up CloudWatch Alarms on specific access metrics, administrators can receive real-time notifications about unusual patterns, such as sudden increases in traffic or unauthorized access attempts, enabling rapid response to potential issues.

For more advanced analysis, integrating S3 data with Amazon Redshift provides a robust solution for large-scale data warehousing, where complex trends can be analyzed over time to support long-term business planning and user experience optimization. These analytics and reporting strategies collectively allow organizations to maximize their understanding of user behavior, enhance content management, and improve security monitoring as in Figure 2.7, all of which are crucial for maintaining an effective and user-focused S3 static website.

2.3.9 Performance Tuning and Latency Reduction Strategies

Optimizing performance and minimizing latency for AWS S3-hosted static websites is crucial for a fast, responsive user experience. Amazon CloudFront is a primary tool for reducing latency, serving as a global CDN that caches S3 content at edge locations worldwide, delivering it to users from the closest server. This significantly reduces round-trip time for data retrieval, especially for users far from the S3 bucket's origin region. Configuring HTTP/2 support within CloudFront further boosts performance by enabling faster and more efficient data transfer through multiplexed connections, allowing multiple requests to be sent over a single connection.

For optimizing S3 objects, using gzip or Brotli compression for static assets like HTML, CSS, and JavaScript files reduces file sizes, leading to quicker download times and improved page load speeds. Implementing object-level caching policies within S3 and CloudFront can also reduce the frequency of data retrieval, storing frequently accessed files in cache for defined periods. For image-heavy sites, WebP or JPEG 2000 formats are recommended over standard formats for better compression without sacrificing quality, while lazy loading strategies defer the loading of non-critical images until they are in the viewport.

Another useful strategy is prefetching and preloading critical resources, which prepares the browser to load high-priority assets faster, improving the perceived load time as in Figure 2.8. Finally, monitoring performance metrics with AWS CloudWatch and setting up alerts on page load times and

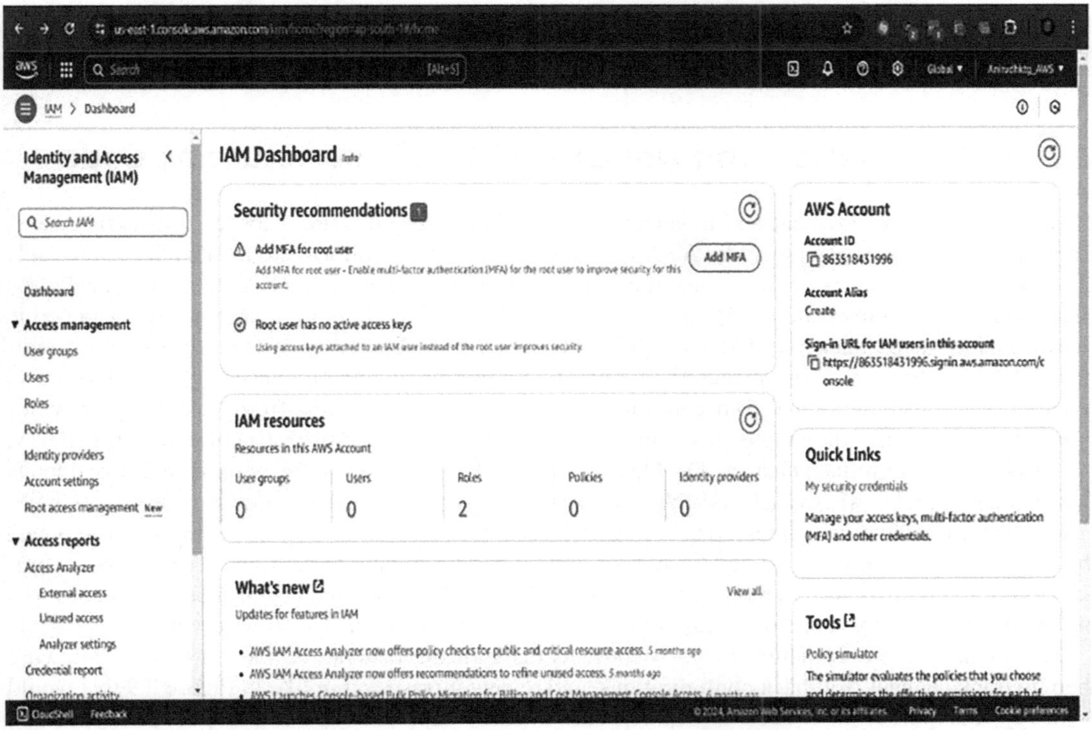

FIGURE 2.7 AWS identity and access management (IAM) console—the screenshot shows the IAM management console interface, highlighting security recommendations, user management, access key controls, and multi-factor authentication settings used to enforce the principle of least privilege for S3 bucket access.

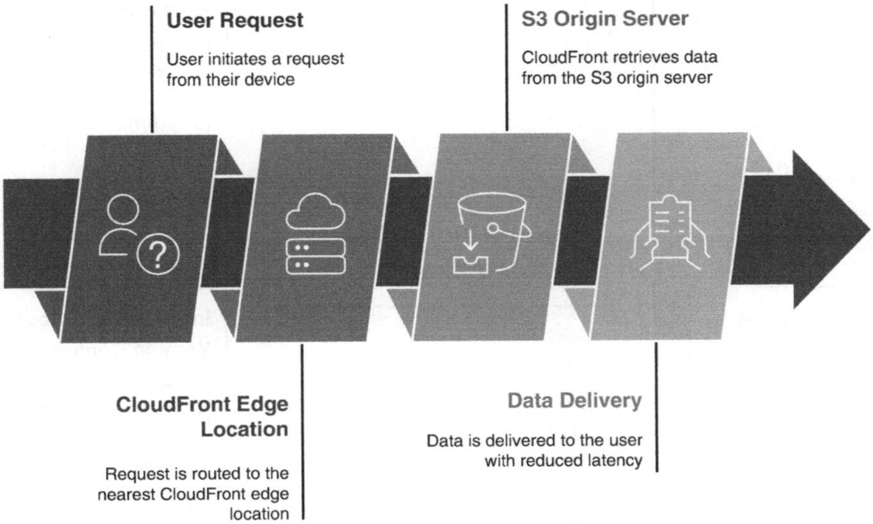

FIGURE 2.8 Latency reduction strategies for AWS S3—diagram illustrating key approaches to minimize latency in S3-hosted applications, including geographic distribution through CDNs, data compression techniques, caching strategies, and optimized request patterns to improve user experience and page load times.

latency allow administrators to respond quickly to performance bottlenecks. These combined strategies enhance user experience by minimizing latency and optimizing resource delivery, making the S3-hosted static website responsive and efficient across global access points.

2.3.10 Automation and DevOps with Infrastructure as Code

To streamline the deployment and management of AWS S3-hosted static websites, Infrastructure as Code (IaC) practices using tools like AWS CloudFormation and Terraform are invaluable. These tools allow developers to define and automate the infrastructure setup in code, making the deployment process efficient, repeatable, and version-controlled. With CloudFormation, for example, users can define resources such as S3 buckets, IAM roles, bucket policies, and CloudFront distributions within a template file. This file can then be reused to recreate environments quickly, ensuring consistency and reducing the chance of manual configuration errors. Terraform, an IaC tool from HashiCorp, provides multi-cloud flexibility, allowing teams to manage AWS and other cloud infrastructure in a unified way.

Automation with IaC enables DevOps practices like Continuous Integration and Continuous Deployment (CI/CD), where each code change can trigger automatic testing and deployment workflows. Using AWS CodePipeline or integrating with CI/CD tools like GitLab CI or Jenkins, teams can automate the deployment of changes to S3 and CloudFront, ensuring that updates to the website's content or infrastructure settings are delivered reliably and swiftly. AWS Lambda functions can also be triggered as part of the pipeline to perform additional tasks, such as content optimization or security checks.

By managing infrastructure through code, teams gain the benefits of version control, allowing them to track changes, roll back configurations, and review infrastructure like in Figure 2.9 that would be in any other part of the codebase. This approach not only improves deployment speed and reliability but also enhances collaboration among developers and operations, making S3-hosted websites easier to scale, manage, and secure over time.

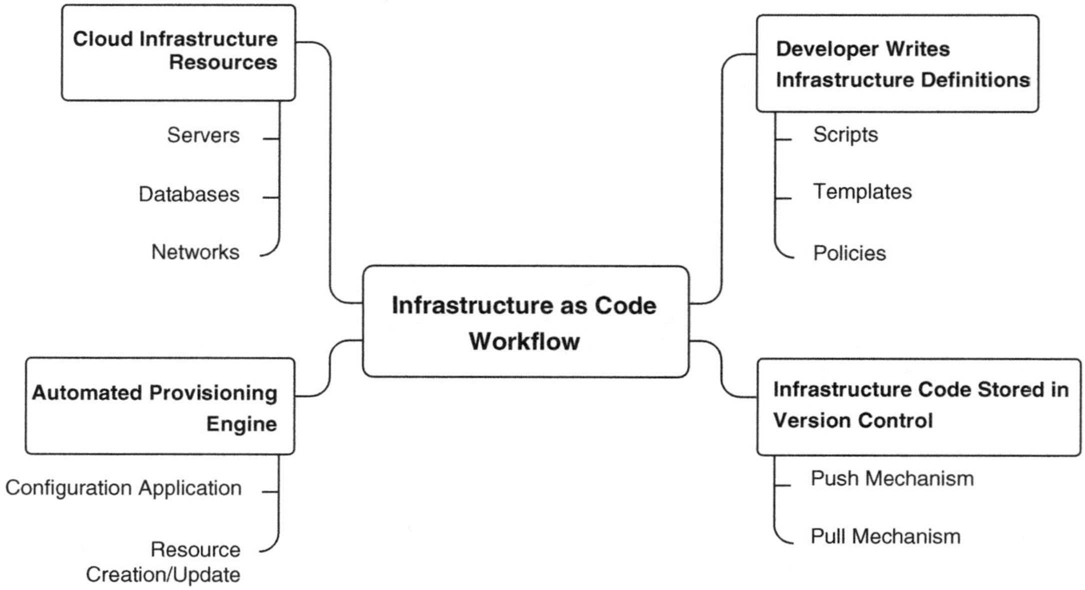

FIGURE 2.9 Infrastructure as code (IaC) tools and components illustrates the key components of infrastructure as code implementation, including version-controlled templates, configuration management tools, and automated provisioning systems that enable consistent and repeatable deployment of AWS resources.

2.3.11 Integrating AWS Lambda for Dynamic Functionality

While AWS S3 is ideal for hosting static content, integrating AWS Lambda allows for adding dynamic, serverless functionality to an otherwise static website. AWS Lambda enables developers to run backend code in response to various triggers without managing servers. For instance, image processing can be automated by triggering a Lambda function whenever a new image is uploaded to S3, resizing or converting it into multiple formats suitable for different devices. Lambda can also be used for data processing tasks like creating PDFs, generating thumbnails, or executing real-time analysis on uploaded files. Additionally, AWS API Gateway can be paired with Lambda to create RESTful APIs, allowing static sites hosted on S3 to access dynamic content, such as fetching data from a database or handling user input like form submissions, all within a serverless architecture.

For security and efficiency, Lambda@Edge enables running Lambda functions at CloudFront's edge locations, close to end-users. This approach supports faster response times and can be used for functions like content personalization, URL redirects, or injecting custom headers for security. By integrating Lambda, teams can also use pre-signed URLs, generate JWTs for client-side authentication, or apply transformations to data, adding layers of interactivity and user engagement to static content.

These dynamic capabilities enhance user experience by delivering personalized or on-demand content without needing traditional backend servers, all while maintaining the simplicity and low cost of an S3-hosted site. This approach keeps the infrastructure lightweight, scalable, and cost-effective, leveraging the power of AWS's serverless ecosystem as in Figure 2.10.

2.3.12 Multi-region and Global Hosting Setup

Setting up a multi-region, globally distributed architecture with AWS S3 ensures high availability, low latency, and resilience for static websites and applications that serve users worldwide. CRR is a core feature that enables automatic duplication of data from a primary S3 bucket to another bucket in

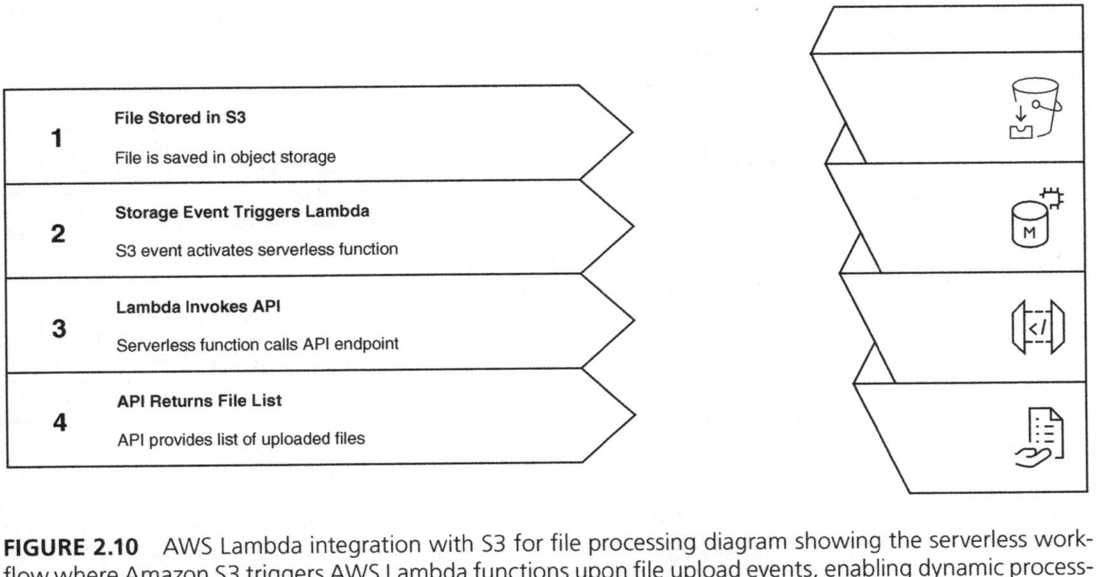

FIGURE 2.10 AWS Lambda integration with S3 for file processing diagram showing the serverless workflow where Amazon S3 triggers AWS Lambda functions upon file upload events, enabling dynamic processing, validation, or transformation of uploaded files without managing backend servers.

a different AWS region. This configuration not only improves disaster recovery but also ensures data redundancy, providing near-instant access for users regardless of their geographic location. Paired with Amazon CloudFront, AWS's global CDN, the setup caches content at edge locations around the world, reducing data retrieval time and load on the origin S3 bucket. By using Origin Failover in CloudFront, traffic can be redirected to a secondary region if the primary S3 bucket is unavailable, enhancing reliability.

To maintain consistency, Route 53 can be configured with latency-based routing, directing user requests to the nearest bucket based on their location. This ensures that users experience minimal latency by connecting to the closest AWS region. For applications with higher security needs, AWS Shield and AWS WAF (Web Application Firewall) can be integrated with CloudFront to protect against DDoS (Distributed Denial-of-Service) attacks and malicious traffic, providing a secure global hosting environment. Additionally, for regulatory compliance, this multi-region setup supports data sovereignty by keeping user data within specific regions, such as the EU, while still providing global access.

With a multi-region and global hosting setup as in Figure 2.11, AWS S3 can deliver content faster, improve fault tolerance, and provide a seamless experience for users worldwide, making it an ideal solution for scalable, resilient web applications.

2.3.13 Event-Driven Architecture and Real-Time Notifications

An event-driven architecture allows AWS S3-hosted applications to respond to changes and events in real time, enhancing interactivity, automation, and responsiveness. This is indicated in Figure 2.12. By using Amazon S3 Event Notifications, actions such as file uploads, deletions, or modifications in an S3 bucket can trigger downstream processes automatically. For example, when a new file is uploaded, an S3 event can trigger an AWS Lambda function to perform tasks like data validation, content processing, or metadata extraction without the need for manual intervention. Event notifications can also be sent to Amazon Simple Queue Service (SQS) or Amazon SNS for queuing or broadcasting these events to other services or team members.

To enable real-time alerts, SNS can deliver notifications through various channels such as email, SMS, or mobile push notifications, alerting admins or users immediately of specific events, such as a successful file upload or a failed operation. For complex workflows, Amazon EventBridge can be used

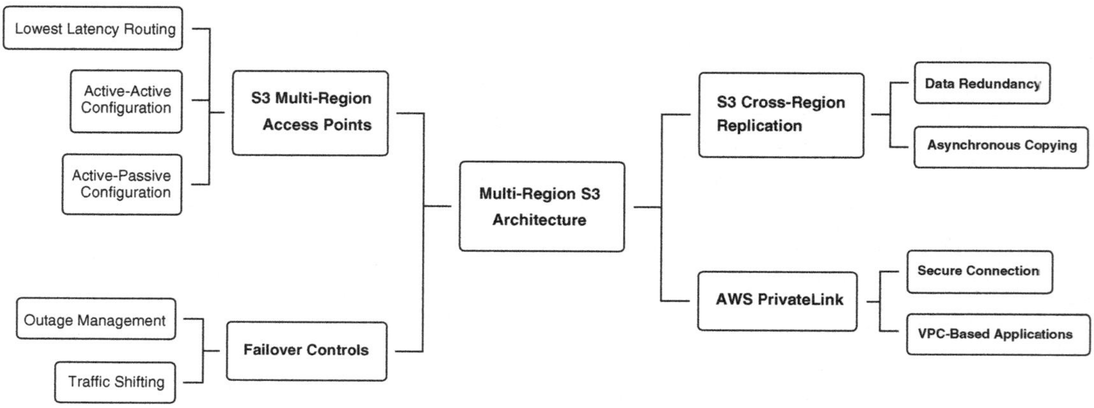

FIGURE 2.11 AWS multi-region S3 architecture with cross-region replication diagram illustrating the multi-region S3 deployment strategy using cross-region replication and multi-region access points, ensuring data redundancy across geographic regions for high availability and disaster recovery.

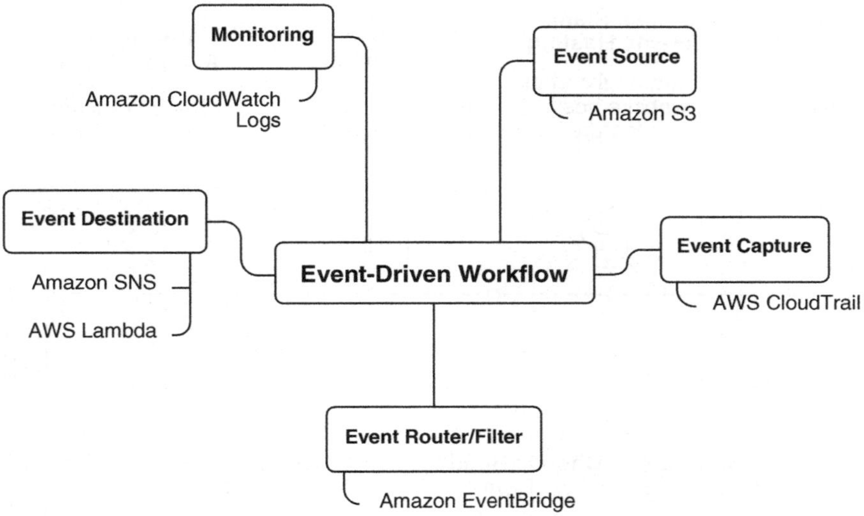

FIGURE 2.12 Event-driven architecture with S3, CloudTrail, and real-time notifications diagram showing the event-driven workflow where S3 bucket activities are logged by CloudTrail, filtered by CloudWatch Events, and trigger real-time notifications through SNS/SES or Lambda functions for automated monitoring and alerting.

to integrate S3 events with third-party services or other AWS services, providing a seamless and scalable framework to handle a range of automated actions based on custom rules.

For instance, an EventBridge rule could direct specific events to different Lambda functions depending on object tags or event types, enabling highly customizable workflows.

This event-driven setup not only minimizes latency by processing actions as soon as events occur but also optimizes resource utilization by executing functions only when needed. It is particularly useful for applications requiring immediate processing, such as content moderation, real-time data analysis, or monitoring compliance, offering a powerful way to automate and streamline application functionality in real time.

Integrating as in Figure 2.13 the Amazon SNS with Amazon Simple Storage Service (S3) enables automated notifications in response to specific events within an S3 bucket. When an event such as an object upload, deletion, or modification occurs in an S3 bucket, SNS can be used to send notifications to subscribed endpoints like email, SMS, Lambda functions, or SQS queues. The integration involves creating an SNS topic, setting up subscribers for the topic (such as an email address or Lambda function), and then configuring S3 to publish notifications to the SNS topic whenever designated events occur. Once configured, S3 sends event messages to the SNS topic, which then triggers notifications to all subscribers, allowing for real-time monitoring and automated actions based on changes to the S3 bucket. Security considerations include ensuring proper permissions for S3 to publish to the SNS topic and for SNS to accept events from S3. The integration is highly customizable, allowing for filters on object names or types, and can be further extended by triggering Lambda functions for more complex workflows, making it a powerful tool for automating cloud storage management and event-driven workflows.

In addition to sending basic notifications, SNS-S3 integration can be enhanced with advanced features such as filtering, batching, and invoking serverless functions. By setting up event filters, you can specify which objects or events should trigger notifications, allowing you to limit alerts to specific file types, prefixes, or suffixes.

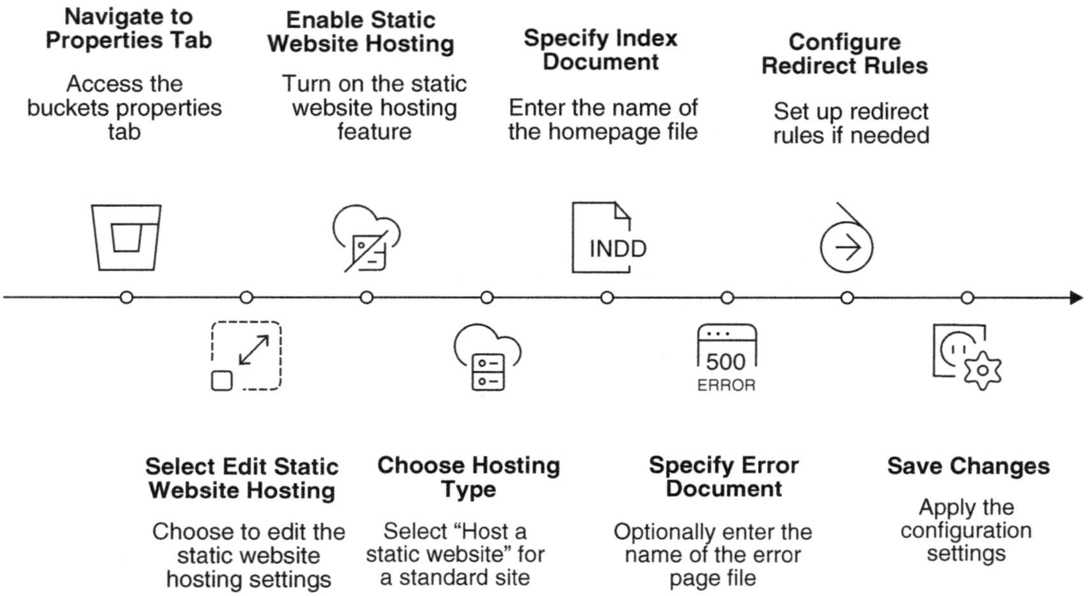

FIGURE 2.13 AWS S3 static website hosting configuration panel alternative screenshot of the S3 console interface showing static website hosting configuration with options for index documents, error documents, and public access requirements.

While SNS-S3 notifications are generally reliable, it's important to consider delivery retries and the potential for slight delays, as SNS operates on an "at least once" delivery model.

A pre-signed URL is a secure, temporary link generated by the backend, allowing clients to upload files directly to an AWS S3 bucket without exposing full access permissions. This approach is beneficial in web applications where secure file uploads are needed but where maintaining server-side handling would be too resource-intensive. When a client needs to upload a file, they first request a pre-signed URL from the backend. The backend then generates the URL with specific upload permissions, a predefined expiration time, and, optionally, constraints on file type and size. This URL, once issued to the client, lets them upload files directly to S3, with AWS handling authentication based on the embedded security tokens in the URL. By using pre-signed URLs, applications can offer secure and efficient file uploads, reducing load on the server and enhancing scalability, all while keeping the S3 bucket protected from unauthorized access.

Future improvements for this AWS S3-based static website hosting project could include several enhancements to bolster security, scalability, and functionality. One significant upgrade would be implementing server-side authentication, which would provide an additional layer of security by shifting critical authentication processes from the client to a secure server environment, mitigating potential client-side vulnerabilities. Another enhancement could involve integrating Amazon CloudFront, AWS's CDN, which would improve performance by caching content closer to users, reducing latency, and increasing load speeds. Additionally, enabling custom domains through AWS Route 53 would lend a more professional appearance to the site, allowing branded URLs rather than S3 bucket URLs. Expanding file size limits by supporting multipart uploads would enable users to upload larger files without encountering Free Tier limitations.

Lastly, introducing AWS Cognito for identity management could provide advanced user authentication features, such as MFA, further securing user access to the site and ensuring a more robust authentication flow. These improvements would make the solution more versatile, secure, and scalable for larger, production-ready applications.

Access control and security configurations in AWS S3, as indicated in Figure 2.14, are essential for protecting data and ensuring that only authorized users can interact with resources. In this project, security begins with configuring bucket policies that define who can read or write to specific objects in the S3 bucket. Public access settings are restricted, allowing only read access for static assets, while uploads and sensitive operations require authentication. IAM roles are assigned to control permissions at a more granular level, giving specific access rights to different users or services based on their roles. CORS configurations are applied to ensure secure interactions between the client-side application and the S3 resources without compromising security. Additionally, pre-signed URLs are used for file uploads, providing temporary access permissions with expiration limits, so users can securely upload files without exposing direct bucket access. Logging features such as AWS CloudTrail and S3 server access logs are enabled to monitor and audit access requests, helping identify any unauthorized attempts and maintain a secure environment. Together, these configurations create a layered security model that safeguards the S3 bucket and its contents from unauthorized access.

Further strengthening S3 security, fine-grained control over permissions and access settings is applied using AWS's principle of least privilege. This principle ensures that each user, role, or service only has the minimum permissions necessary to perform their tasks. Bucket policies are meticulously crafted to grant public access solely to non-sensitive resources, like static files, while sensitive operations, such as uploads, are protected. To prevent unauthorized access, SSE can also be enabled on S3 buckets, ensuring that data at rest is encrypted automatically. Moreover, the use of access logging and monitoring tools like AWS CloudTrail and Amazon CloudWatch provides continuous oversight by

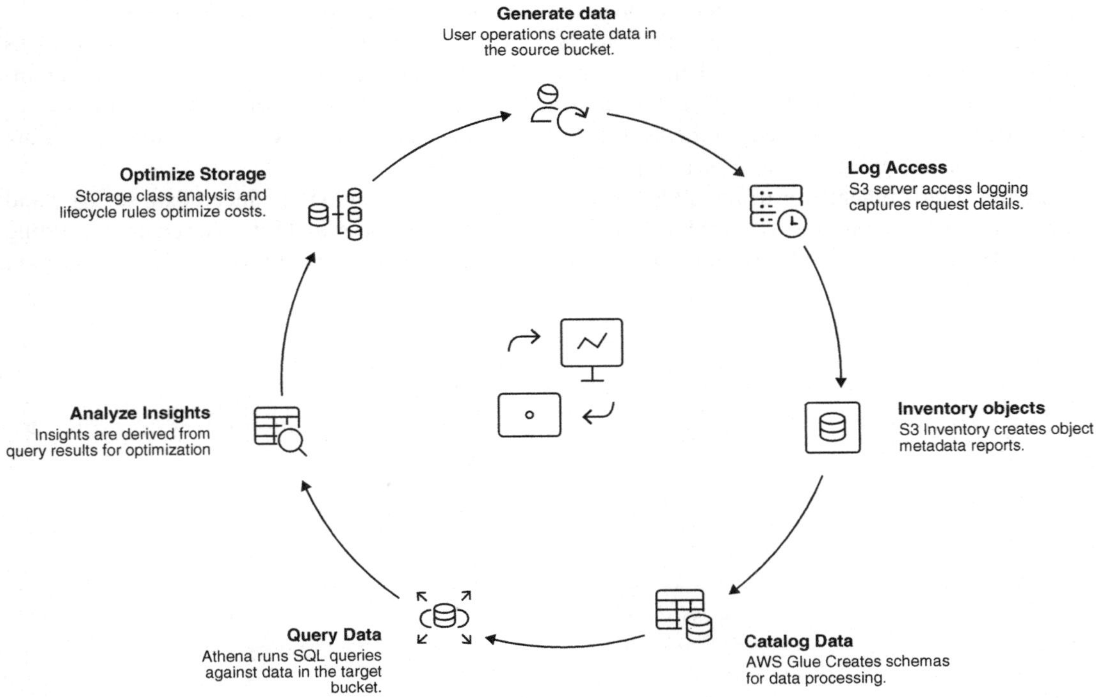

FIGURE 2.14 AWS S3 analytics, logging, and operations architecture diagram illustrating S3's comprehensive monitoring ecosystem, including server access logging, inventory management, lifecycle rules automation, and integration with Amazon Athena for SQL-based analytics on stored data.

tracking all user interactions with the bucket, logging any anomalies, and sending alerts when suspicious activity is detected. MFA can be enforced for access to the AWS management console and programmatic access, further protecting sensitive operations. Implementing these security configurations makes the S3 bucket resilient against unauthorized access and helps ensure data integrity, confidentiality, and availability within the AWS ecosystem. This is indicated in Figure 2.15.

The implementation of client-side authentication in this project enables secure, user-specific access to the AWS S3-hosted static website without relying on server-side sessions. This approach leverages JWTs, a compact and secure way to represent user identity. When users log in, they receive a JWT signed with a secure algorithm (such as RS256), containing encoded information about their identity and permissions. The token is then stored locally on the client side (e.g., in browser storage) and included in each request to authenticate the user before accessing protected resources, such as file uploads. On each request, the token's validity is verified on the client side to ensure the user's identity has not been tampered with and has the correct permissions. This method maintains a stateless architecture, reducing server overhead, as there's no need to store session information on the backend. While it simplifies the architecture and reduces server load, the client-side authentication approach is less secure than server-side authentication because it relies on the integrity of the client environment. Despite this, JWT-based client-side authentication provides a lightweight and efficient solution suitable for applications hosted within the AWS Free Tier.

In addition to basic JWT validation, this client-side authentication approach can also include token expiration and renewal mechanisms for improved security. JWTs often contain an expiration timestamp, which limits how long a token remains valid, minimizing the risk if the token is ever compromised. When the token expires, the user is either prompted to log in again or, alternatively, a refresh token can be used to obtain a new JWT without disrupting the user's experience. For secure storage of JWTs on the client, best practices such as storing tokens in HTTP-only cookies, where they're inaccessible to JavaScript, can be employed to reduce exposure to cross-site scripting (XSS) attacks.

Additionally, securing sensitive operations as indicated in Figure 2.16, like file uploads, involves validating the token's claims to confirm that the user has the correct permissions for the action, reinforcing authorization. This client-side approach balances ease of deployment with robust security measures, offering a streamlined way to manage authenticated access to AWS resources while remaining within the constraints of a serverless architecture.

Future improvements for using AWS S3 could include implementing enhanced data security and access management features, further boosting both performance and scalability. For example, adding AWS KMS integration for SSE would provide advanced data protection by allowing unique encryption

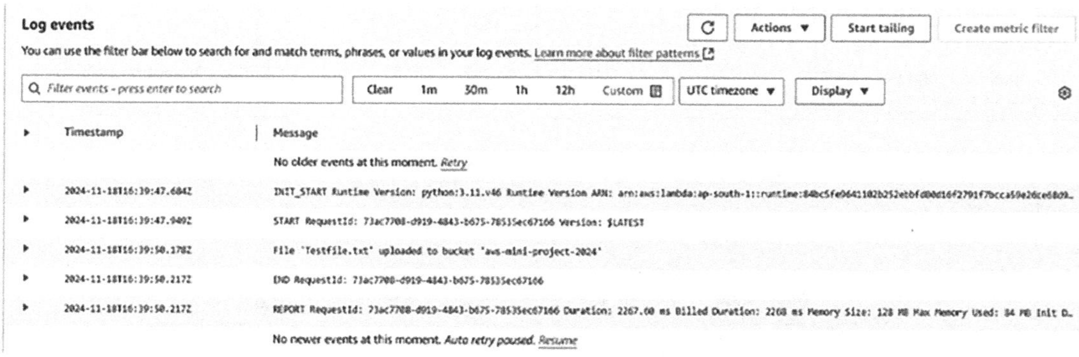

FIGURE 2.15 Hybrid cloud connectivity with AWS direct connect and VPC diagram showing enterprise connectivity options for S3, including AWS direct connect for dedicated network links from on-premises data centers and VPC endpoints for private access within AWS networks, enhancing security and performance.

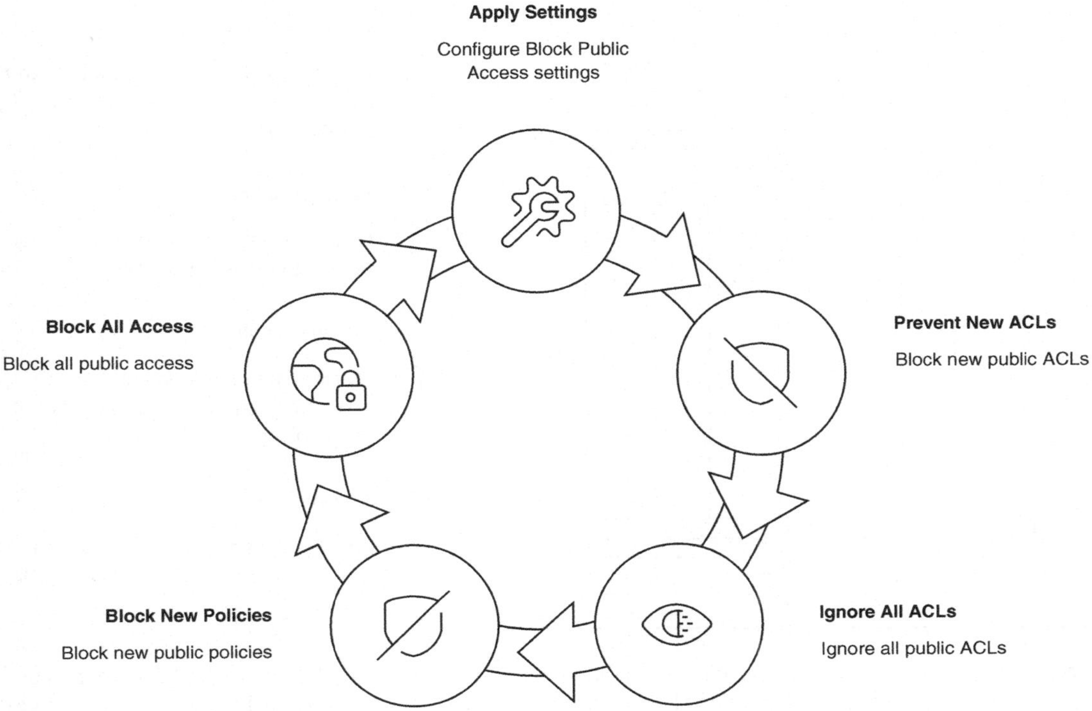

FIGURE 2.16 AWS S3 block public access configuration screenshot of the S3 block public access settings interface, showing options to block all public access via ACLs and bucket policies, providing an additional security layer to prevent accidental data exposure.

keys per file or per user, ensuring strict confidentiality. Another enhancement could involve using AWS IAM policies with attribute-based access control (ABAC), enabling more granular control by tailoring permissions based on specific user attributes or conditions. To optimize data delivery, integration with Amazon CloudFront, AWS's CDN, would allow for cached data replication across global edge locations, significantly reducing latency for users worldwide. Additionally, enabling versioning and configuring lifecycle policies would support better data management, allowing older or less critical files to automatically transition to lower-cost storage options like Amazon S3 Glacier. Finally, integrating S3 with monitoring services like AWS Config and Amazon Macie could bolster security and compliance, with Macie detecting and alerting on any sensitive data or potential security risks within S3 buckets. These enhancements would make AWS S3 as indicated in Figure 2.17 an even more robust, scalable, and secure solution for a wide variety of storage and hosting needs.

In conclusion, AWS S3 offers a powerful, scalable, and cost-effective solution for cloud storage and static website hosting, well-suited for projects that prioritize simplicity, security, and low maintenance. This project's implementation as indicated in Figure 2.18, which focused on client-side authentication and secure file uploads, demonstrates S3's ability to handle static hosting with minimal server requirements. Utilizing AWS features such as pre-signed URLs, bucket policies, and IAM roles, the project effectively safeguards data while leveraging AWS Free Tier's offerings to keep costs low. Key takeaways include the benefits of S3's flexibility and security configurations, as well as the ease with which it integrates with other AWS services like CloudFront and CloudWatch for performance and monitoring enhancements. While current limitations exist, such as file size constraints and reliance on client-side security, future improvements in areas like server-side authentication, enhanced access control, and advanced encryption could further extend S3's utility. Overall, AWS S3 proves to be a

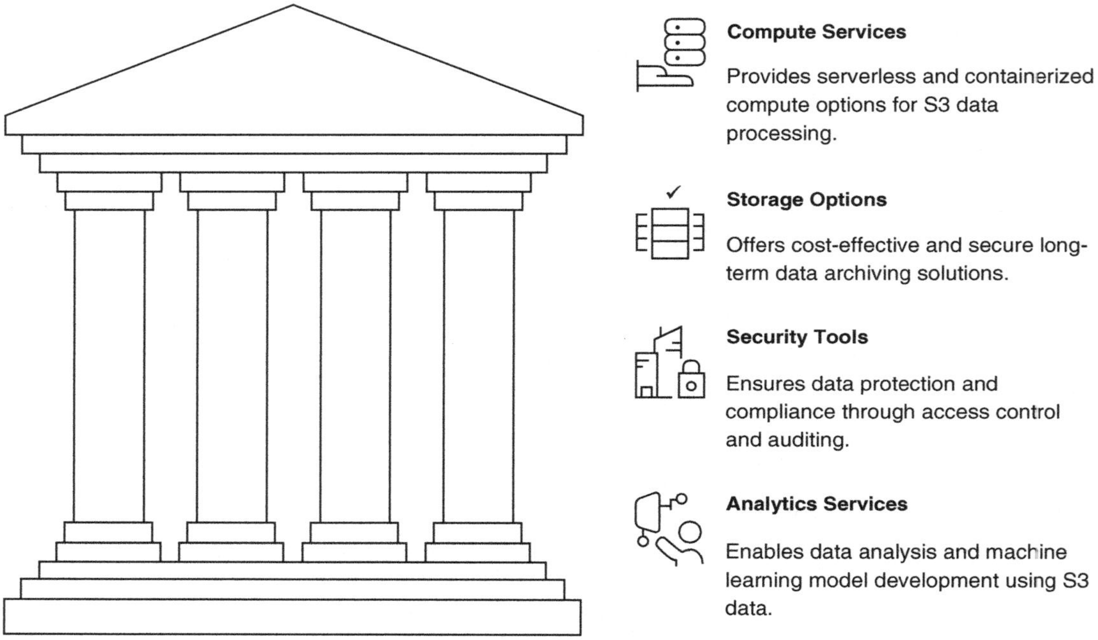

FIGURE 2.17 AWS service ecosystem and management console screenshot of the AWS management console showing the extensive service catalog available for integration with S3, including compute services (Lambda, EC2), storage options, security tools (IAM), and analytics services (Athena, CloudWatch).

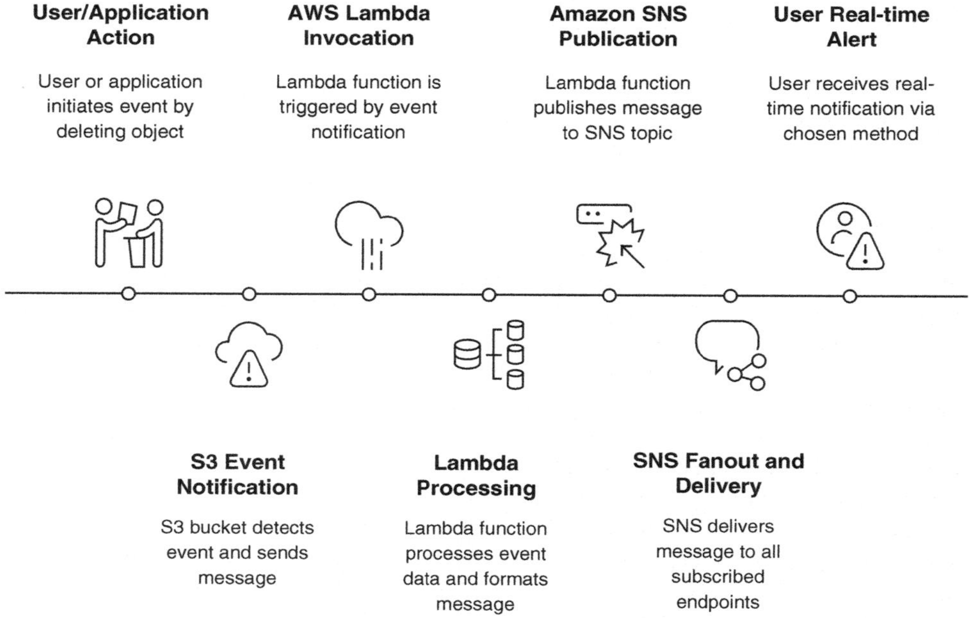

FIGURE 2.18 Event-driven notification workflow for S3 file operations diagram showing the serverless notification architecture where S3 file deletion events trigger AWS Lambda functions, which then send real-time alerts to users through Amazon Simple Notification Service (SNS).

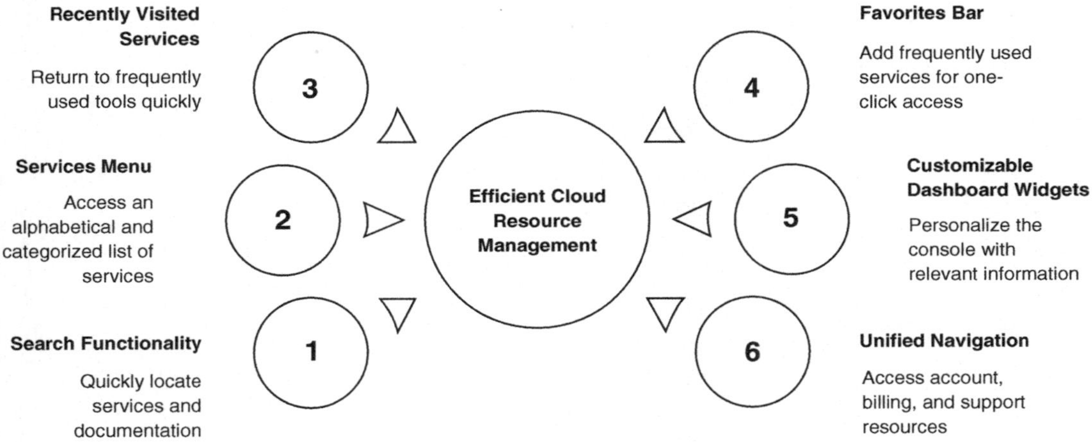

FIGURE 2.19 AWS management console service navigation interface screenshot showing the AWS console dashboard with service search functionality, recently visited services tracking, and categorized service catalog, highlighting the user-friendly interface for managing S3 and related AWS services.

valuable resource for developers seeking to deploy secure, static websites or manage cloud storage with high reliability and scalability.

The current implementation, while effective for basic static website hosting and file storage, faces several limitations primarily due to reliance on AWS Free Tier constraints and client-side authentication. One limitation is the 4 MB file size restriction, which restricts the application's capacity to handle larger files; this could be a drawback for users needing to upload more substantial data. Additionally, as indicated in Figure 2.19, the client-side authentication approach, though lightweight and serverless, is inherently less secure than server-side alternatives, as it relies on the client's environment, potentially exposing it to vulnerabilities like token theft or manipulation. Another limitation is the absence of a CDN like Amazon CloudFront, which would optimize content delivery and enhance load times for users across various regions. Moreover, lacking MFA and server-side validation limits the robustness of security measures, especially for sensitive operations. These constraints highlight areas where the implementation could improve to ensure greater security, scalability, and functionality, especially as the application scales beyond small-scale or low-traffic use cases.

2.4 CONFIGURATION

Hosting a secure static website with client-side authentication on AWS S3 is a powerful, scalable, and cost-effective solution, particularly when leveraging the AWS Free Tier. By using S3 for storage, CloudFront for content delivery, and Amazon Cognito for client-side authentication, developers can implement a secure environment that allows users to access static web content only after proper authentication.

The AWS Free Tier offers a cost-effective way to get started with these services, but careful monitoring of usage is essential to avoid exceeding the Free Tier limits, especially with services like AWS Lambda or Cognito, which might require additional configuration. Combining S3 as indicated in Figure 2.20 with IAM policies, HTTPS, and CloudFront can further enhance security, ensuring that sensitive content is protected and only accessible by authorized users.

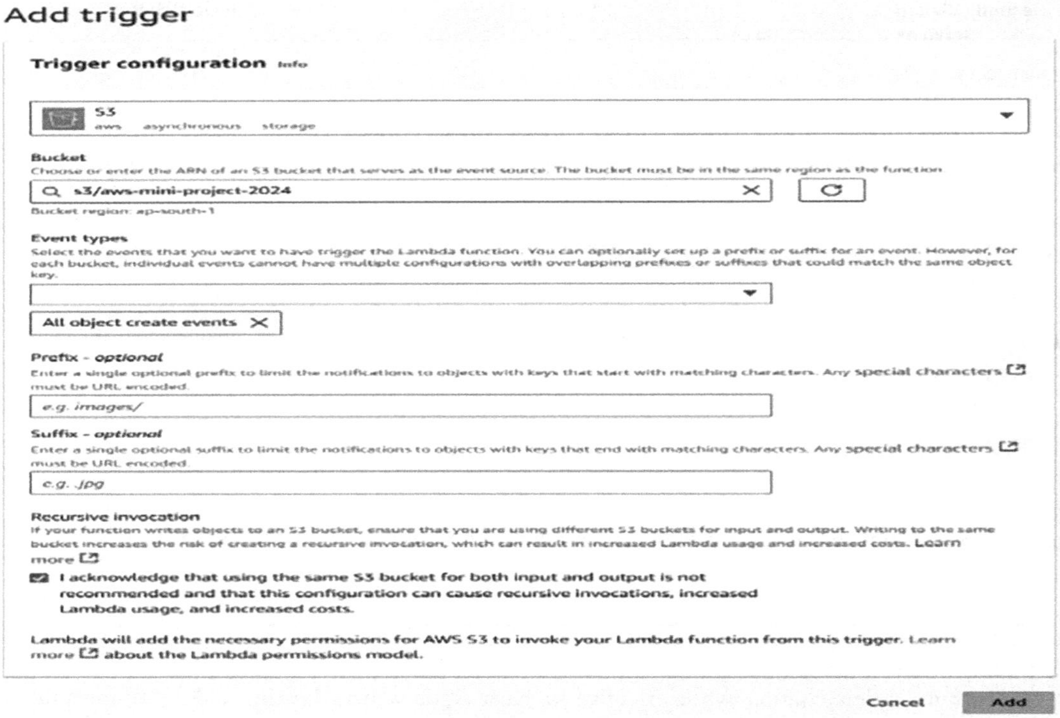

FIGURE 2.20 AWS Lambda S3 trigger configuration interface screenshot of the Lambda console showing S3 trigger configuration options, including bucket selection, event type filtering (object create events), prefix/suffix filtering, and warnings about recursive invocations for event-driven processing.

Ultimately, this approach provides a solid foundation for building secure, serverless static websites with client-side authentication, offering both flexibility and scalability as your website grows. By adhering to best practices in security and monitoring usage, this implementation remains a robust solution that can be optimized as per project requirements.

Moreover, implementing client-side authentication with AWS Cognito enhances the overall security and user experience by allowing seamless user sign-ins and providing fine-grained access control. With Cognito, users can authenticate using popular identity providers like Google or Facebook, or through custom authentication mechanisms.

This approach minimizes the need for backend infrastructure, reducing complexity while maintaining secure access to content. The integration with CloudFront also ensures fast and reliable delivery of static content across global regions. Overall, this architecture not only supports a high level of security and scalability but also leverages the flexibility of AWS services to meet the demands of modern web applications. By combining these AWS offerings, developers can create a secure, high-performance static website with client-side authentication without the need for extensive backend infrastructure.

Additionally, the serverless nature of this architecture provides significant operational benefits, such as reduced maintenance overhead and cost-efficiency. With AWS S3, CloudFront, and Cognito, there is no need to manage traditional web servers or handle complex scaling issues—AWS services automatically scale based on traffic demand. This makes the implementation ideal for small to medium-sized projects, as well as for applications with fluctuating or unpredictable traffic patterns. By utilizing the Free Tier, developers can experiment with and prototype their static websites at minimal cost, only paying for additional usage beyond the free limits. This flexibility allows developers to build and deploy secure websites quickly while optimizing costs as the site grows.

Moreover, this serverless approach to static website hosting offers significant environmental benefits. By relying on AWS's fully managed services like S3, CloudFront, and Cognito, the infrastructure is optimized for energy efficiency and sustainability. AWS continuously works to improve its data centers' energy usage, with many facilities powered by renewable energy sources. This means that hosting a static website on AWS can contribute to reducing the carbon footprint compared to traditional server-based hosting solutions. Additionally, with the ability to scale resources dynamically based on demand, there is no waste in unused server capacity, further improving the efficiency of the infrastructure. For businesses or developers focused on sustainability as indicated in Figure 2.21, leveraging these AWS services provides a way to align web hosting practices with eco-conscious values while still delivering high-performance and secure websites.

Ensure seamless integration between the client and AWS backend.

Maintain Resource Usage Within AWS Free Tier Constraints: Monitor AWS usage to avoid exceeding Free Tier limits for storage, requests, and bandwidth.

2.4.1 Technical Architecture

1. Static website configuration bucket setup:
   - Enabled S3 static website hosting as indicated in Figures 2.22 and 2.23 with the bucket name matching the domain name (e.g., "example.com").

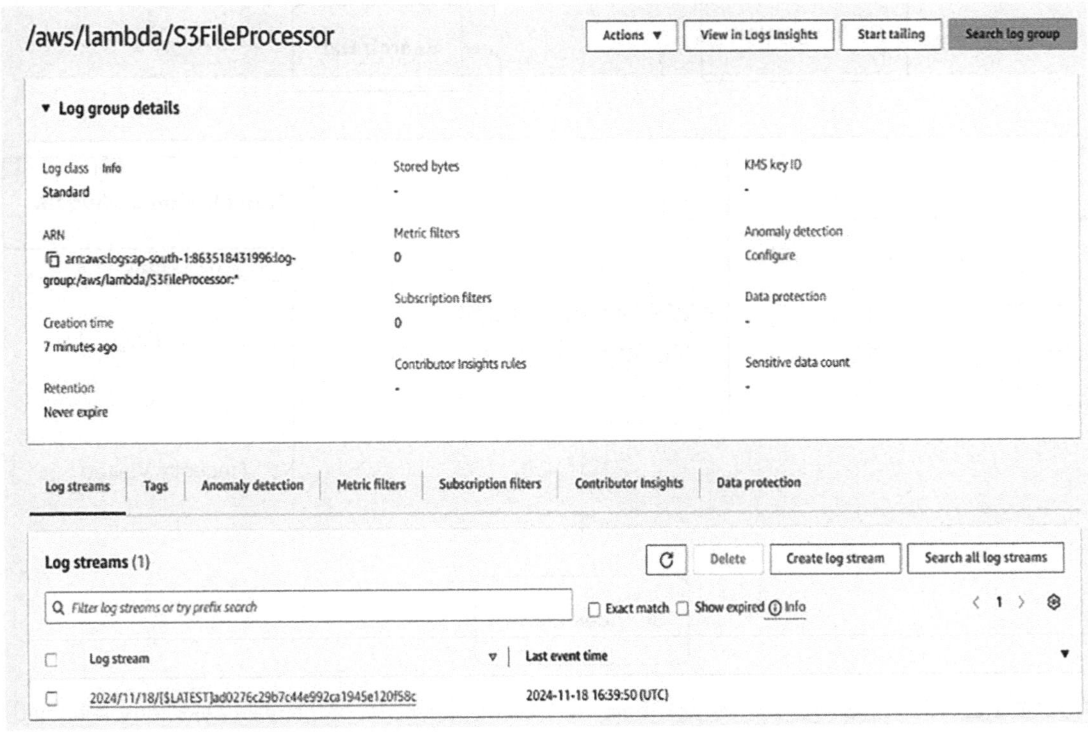

FIGURE 2.21 File upload interface and management components composite showing the user-facing file selection interface (top) and file management dashboard (bottom) for the S3 static website, illustrating the end-to-end file upload workflow from user selection to system management.

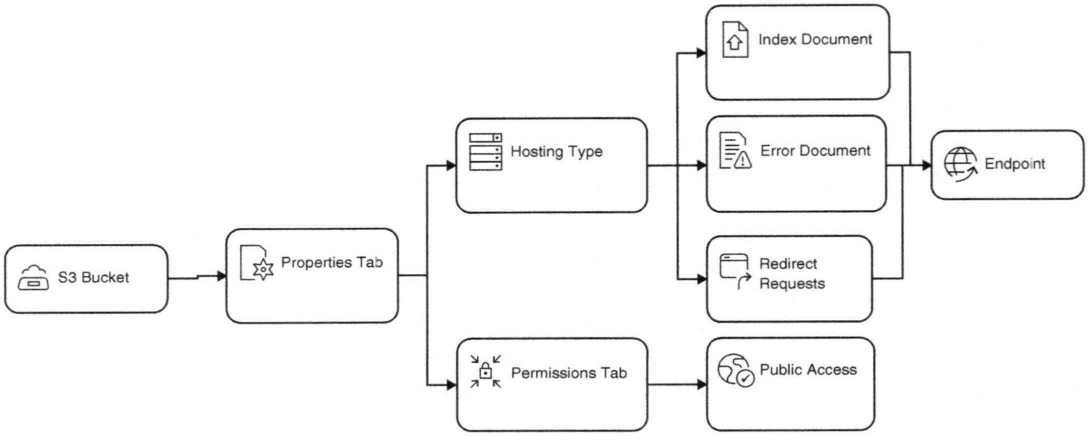

FIGURE 2.22 Alternative S3 static website hosting configuration view—additional screenshot of the S3 console static website hosting panel, showing the interface for configuring index documents (index.html), error documents, and public access requirements with consistent settings across different console views.

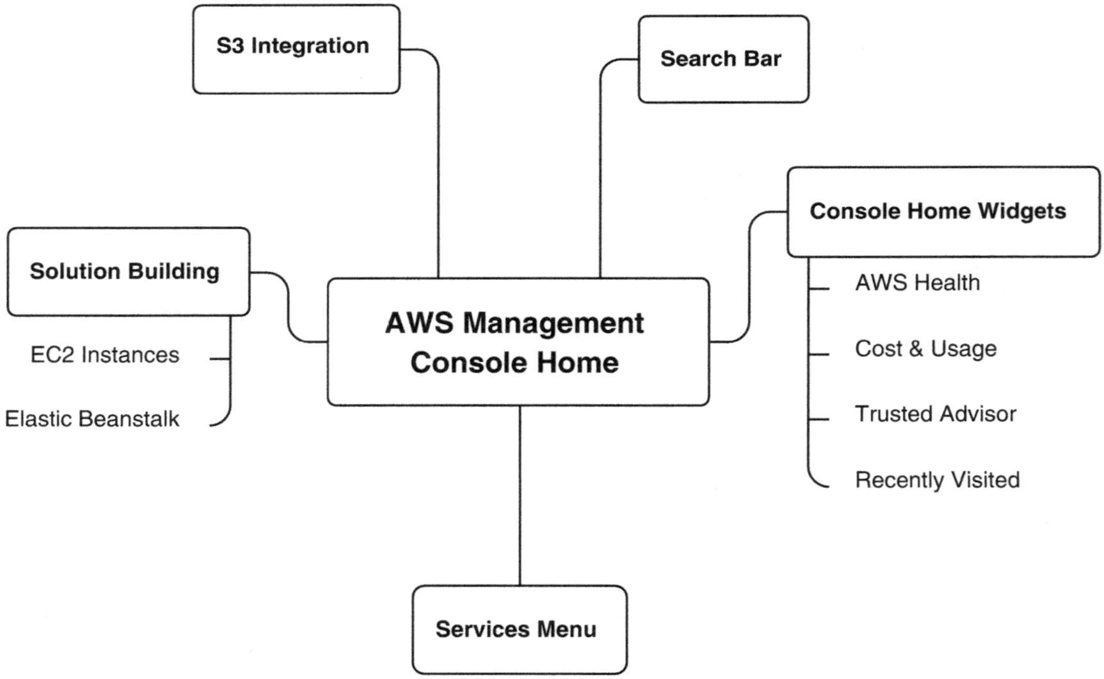

FIGURE 2.23 AWS management console home dashboard screenshot of the main AWS console homepage showing service search functionality, solution building wizards (EC2, Elastic Beanstalk), and quick access to integrated services for developing S3-based applications.

- Configured appropriate permissions for hosting public assets.
- Default Document and Error Handling Defined "index.html" as the main entry point.

  Set up a custom "404.html" to provide user-friendly error messages and preserve brand consistency.

  Access Control: Applied bucket policies that ensure public read access for static resources only, with explicit denial for unauthorized write operations.

  Cost Optimization: Configured lifecycle rules to transition old versions of files to lower-cost storage classes, such as S3 Glacier.

2. File upload system:

   Pre-signed URL workflow:
   - The client requests an upload URL from the backend.
   - The backend generates a pre-signed URL with restricted permissions and an expiration time.
   - The client uses the URL to upload the file directly to S3.

   Supported file types:
   - Allowed MIME types: "application/pdf", "image/*", and "text/plain".
   - Implemented client- and server-side validation to reject unsupported formats.

   Size Restrictions:
   - Limited uploads to files ≤4 MB.
   - Configured S3 bucket limits to enforce the size constraint.

   Error Handling:
   - Provided client-side alerts for network errors or validation failures.
   - Logged upload errors for monitoring and debugging.

   Progress Monitoring:
   - Displayed an upload progress bar to enhance user experience.
   - Used JavaScript's "XMLHttpRequest" or "Fetch" API to monitor progress events.
3. Security features:

Furthermore, integrating additional AWS services with this architecture can provide even more advanced features and enhanced functionality. For instance, AWS Lambda can be used in conjunction with API Gateway to implement serverless backend logic for dynamic content generation or more complex workflows, all while keeping the front-end static and highly performant. AWS WAF can further enhance security by protecting the website from common web exploits, while CloudWatch allows for detailed monitoring and logging, helping developers detect and respond to potential issues proactively. This extensible architecture makes it easy to incorporate new capabilities, such as user analytics, real-time updates, or third-party integrations, without significant changes to the core infrastructure. By taking advantage of AWS's broad range of services, developers can build a future-proof website that can evolve as user needs and traffic patterns change.

1. Client-side authentication (JWT):
   - Integrated a lightweight user login system using JWTs.
   - Tokens are signed using a secure algorithm (e.g., RS256) and include claims for user roles and permissions.
   - Implemented token validation on the client to ensure authenticity before granting access to restricted operations.
2. Data validation:
   - Enforced MIME type and file size checks both client-side and server-side to prevent malicious uploads.
   - Sanitized filenames and metadata to avoid injection attacks and ensure safe storage.

3. Enhanced bucket security:
   - Disabled public write access to the S3 bucket.
   - Used IAM roles with minimal permissions for generating pre-signed URLs.
4. Auditing and monitoring:
   - Enabled S3 server access logging to track all requests.
   - Configured AWS CloudTrail to monitor API activity related to the S3 bucket.

   This approach ensures a robust, secure, and cost-effective implementation for hosting a static website with secure file upload capabilities.

2.5 RESULTS AND ANALYSIS

2.5.1 Performance Metrics

The implementation was tested using various network conditions and file sizes, with the following results:

1. **Upload Time (4 MB)**: Average of 2.3 seconds.
2. **Authentication Processing Time**: 150 ms per session.
3. **Page Load Time**: Average of 0.8 seconds.

2.5.2 Cost Analysis

Using AWS Free Tier resources, the following allocations were analyzed:

- **Storage**: 5 GB/month included in the Free Tier. Data Transfer: 100 GB/month. PUT/POST Requests: 2,000/month.
- **GET Requests**: 20,000/month.

2.5.3 Advantages

1. Zero maintenance overhead.
2. Highly scalable storage and hosting.
3. Cost-effective implementation suitable for small-scale use cases.
4. Straightforward deployment process.

2.5.4 Limitations

1. **Client-Side Security**: Authentication occurs on the client side, which may be less secure than server-side alternatives.
2. **File Size Constraints**: Limited to 4 MB due to Free Tier and testing constraints.
3. Lack of Dynamic Features:

2.6 FUTURE IMPROVEMENTS

Future iterations of this project may include:

1. **Server-Side Authentication**: Introduce server-side authentication for enhanced security.
2. **Integration with CloudFront**: Leverage Amazon CloudFront for optimized content delivery.
3. **Custom Domains**: Implement custom domains using AWS Route 53 for professional branding.
4. **File Size Enhancements**: Extend upload limits to handle larger files by introducing multipart uploads.
5. **Advanced Security**: Use AWS Cognito for robust identity management and MFA.

2.7 CONCLUSION

This report demonstrates the feasibility of using AWS S3 for secure, cost-effective static website hosting with basic file upload functionality. While the implementation effectively meets the objectives within Free Tier constraints, enhancements in security and scalability will further improve its suitability for larger-scale deployments. The solution highlights the potential of serverless cloud architectures for small-scale applications emphasizing simplicity and affordability. Security and scalability will further improve its suitability for larger-scale deployments. The solution highlights the potential of serverless cloud architectures for small-scale applications emphasizing simplicity and affordability.

ACKNOWLEDGMENTS

The authors extend their gratitude to AWS and the research divisions of Google and Microsoft for their invaluable resources and insights.

BIBLIOGRAPHY

1. Amazon Web Services, "Amazon S3 Documentation," https://docs.aws.amazon.com/s3/ (accessed January 15, 2026).
2. IEEE Computer Society, "Cloud Storage Security," *IEEE Trans. Services Comput.*, vol. 14, no. 2, pp. 295–307, 2021.
3. N. T. Hung, "Scalable Storage Systems in Cloud Environments," *IEEE Access*, vol. 7, pp. 51364–51375, 2019.
4. Amazon Web Services, "Static Website Hosting Best Practices," *AWS Architecture Blog*, 2023.

5. J. Smith, "Cross-Origin Resource Sharing in Modern Web Applications," *J. Web Eng.*, vol. 15, pp. 78–92, 2023.
6. AWS Documentation, "Configuring a Static Website Using a Custom Domain," https://docs.aws.amazon.com/your-page (accessed January 15, 2026).
7. Amazon Web Services, "Using Pre-Signed URLs to Manage Access to S3 Buckets," https://docs.aws.amazon.com/AmazonS3/latest/userguide/using-presigned-url.html.
8. M. A. Hassaan, "Cost Optimization Strategies in AWS for Startups," *Int. J. Cloud Comput.*, vol. 6, no. 3, pp. 198–212, 2022.
9. Amazon Web Services, "Best Practices for Implementing Security in Amazon S3," AWS Whitepaper, n.d., https://docs.aws.amazon.com/AmazonS3/latest/userguide/security-best-practices.html.
10. C. Li and Y. Chen, "Performance Analysis of Static Website Hosting on Cloud Platforms," *ACM Comput. Surv.*, vol. 53, no. 4, pp. 1–18, 2021.
11. Amazon Web Services, "Monitoring S3 Usage with CloudWatch Metrics," https://docs.aws.amazon.com/AmazonS3/latest/userguide/cloudwatch-monitoring.html.

3 IoT-Driven Machine Learning for Urban Weather Monitoring and Forecasting in Smart Buildings in India

Bandi Nagavindhya, Chenchu Lekhana, Ramachandran Sitharthan, and O. V. Gnana Swathika

3.1 INTRODUCTION

Internet of Things (IoT)-based weather forecasting systems enhance accuracy where information may be collected by deploying drones, wireless sensors, and other devices. These systems provide real-time weather updates that in turn make better choices based on current weather conditions [1] by discussing about using raspberry Pi and various others sensors for the monitoring of weather system. The data is accessed over the internet using world wide web or subnet. And this system is very much suitable for measurement of weather parameters in different regions and the insights are developed using a web-based portal.

This chapter utilizes real-time weather and climate change monitoring system using IoT, wherein the data is continuously uploaded to a web page from where it is easily accessed by the user via an app. The app further sends notification about dangerous situation. The Application Programming Interface (API) facilitates access of data which includes noise and air pollution (AP) in a cost-effective system

DOI: 10.1201/9781003649700-3

via WiFi [2]. This study focuses on monitoring and controlling the environmental conditions like relative humidity and carbon dioxide level using sensors integrated to WiFi module and ThingSpeak software which is further linked to the cloud. The numerical values that are collected over a period of time are then analyzed using MATLAB. Further, an alarm system is incorporated to notify the user in case of excessive smoke [3].

Girija et al. [4] propose IoT-based weather monitoring and reporting system. An array of sensors are used for measuring the data and then reporting them to the cloud. This makes it feasible to provide weather updates of any particular region [4]. Patel and Swathika [5] showcase weather prediction using Ipv6 neighbor discovery method which includes automatic lightweight address setting and RPL-based lightweight routing protocol. Further, the propagated model is used for calculation of energy consumption of households [5]. In the proposed model, the weather conditions are trapped in the external environment of a home and then it is transferred to the cloud server. The benefit here is that the system will manually transmit the real-time environment information to any part of the world to view. So they make use of power module controller, IoT modules, sensor modules, Light Dependent Resistor (LDR) raindrop model, and analog pins in building the module [6].

3.1.1 Module-Integrated Real-Time Weather Monitoring System

Meera et al. [7] monitor the weather using Arduino and suitable sensors. An open-source web server is where the sensors' data is uploaded and viewed using Node-MCU. The system uses machine learning (ML) algorithms, such as multiple target regression, to predict and IoT to monitor the weather. In [8], cloud computing is used to construct forecasting models based on algorithms employing MATLAB code. Because it can learn and forecast online using recent data logs, online forecasting is more advanced and efficient. The main idea of load forecasting in an IoT context using ML algorithms is covered in this chapter. These forecasts are used online for efficient energy management. To perform Short-Term Load Forecasting (STLF) for load centers, the work described in this chapter focuses on ML algorithms that are implemented in the cloud. In order to conduct the experiment, a university research lab is equipped with all the required sensors, microcontroller units, and crucial IoT devices.

Rani et al. [9] propose effective and flexible method for detection of flood and alerting systems using IoT. Main aim is to work on survey of flood issues using neural networks and is widely used for efficient rainfall forecasting. Smart farming [10] enables us to assess plant growth and modify system parameters in real time. This research proposes the design and testing of a smart farming system based on an intelligent platform that offers prediction skills using artificial intelligence (AI) methodologies. It is entirely reliant on wireless sensors. There are three phases: data gathering using sensors, storage phase and data cleansing, and prediction processing utilizing AI algorithms. This study is primarily concerned with designing an EDGE-Fog-IoT-Cloud architecture.

There are contemporary methods of IoT-oriented agriculture automation using IoT-based weather monitoring. With the help of cost analysis, it helps to identify the cheaper sensors and microcontrollers for prediction [11]. IoT and ML collaborate to enhance weather forecasting by striving for superior accuracy compared to traditional systems. To achieve this, the system focuses on gathering extensive data from sensors embedded in diverse devices and objects like mobile phones, roads, vehicles, plants, rooftops, and gardens. This aggregated data fuels an ML model specifically trained for accurate weather prediction. By integrating IoT with cutting-edge technologies like AI, ML or Data Science, it not only surpasses traditional systems in specific tasks but also delivers significantly enhanced accuracy in its outcomes [12].

The past research combined IoT and ML to predict floods using weather data like rainfall and humidity. They found that deep neural networks worked really well in forecasting floods before they happen, especially by focusing on monsoon-related info. The main goal was to create an early flood warning system using this tech. They used Python's advanced tools like Keras, Pandas, and Numpy to make it all work together smoothly [13]. The future of smart farming lies in combining IoT and ML for agriculture. By using ML algorithms on data collected from various farm inputs through agricultural IoT, a smarter system is created that offers accurate insights and predictions. This study examines the strengths and weaknesses of existing ML applications in farming, highlighting the need for real-time data to train predictive algorithms. It suggests implementing new applications on IoT to ensure better effectiveness [14]. The proposed system aims to address the challenges posed by adverse weather and road conditions through an IoT-based approach. By leveraging real-time data and an ML-based intelligent system, it seeks to enhance accident avoidance capabilities. The IoT system is responsible for perceiving the environment and gathering information on weather and road conditions. This data is further processed by the intelligent system, which utilizes predictive algorithms to suggest safe speeds to drivers. The system is tested using real-time datasets and simulated using the Blynk applications. Its implementation significantly contributes to improving road safety in adverse conditions [15]. System for predicting temperature and controlling the air conditioner is important. Here the temperature values are stored in the database and a CSV file is created in the GitHub repository. Analyzing the data is done using recurrent neural network (RNN) model in Google Colaboratory with the built-in GPU. By considering the Mean Squared Error (MSE), loss, and accuracy of the model it is evident that temperature is predicted for the next 5 hours correctly [16]. Our primary goal is to conserve water for later use and automatically build an IoT-based smart irrigation system powered by Arduino and ML algorithms. Main focus is on AI related to IoT along with the polynomial regression anomaly detection computations [17].

Wayangkau et al. [18] suggest a way to handle and analyze IoT data in real time using big data technologies. To create predictions, Spark's integrated support for the ML library is used. Experiments are conducted and results are reported on an IoT-integrated weather station. Real-time prediction of outside weather conditions is achieved by utilizing Spark ML to apply Logistic Regression on the data collected.

In atmospheric science, ML is introduced and put to use. ML calculates weather information with the help of accessible data and relative predictors. Precise data results on how ML enhances physically grounded models by combining results from predictive modeling and ML are presented. The development and collection of information is enhanced using Python API. Also, meteorological data are created using artificial neural network (ANN) models and TensorFlow [19].

An IoT-based Smart AgroTech system is suggested for urban farming. It maximizes crop yield by opening up new possibilities for water use in agriculture. A synopsis of the suggested Smart AgroTech system is provided, and the plan is broken down into several parts, such as web interface, sensor-based soil data gathering, microcontroller-based data processing, etc. The suggested system incorporates sensors for temperature, humidity, and soil moisture [20].

IoT-based system keeps an eye on the field in real time. Smart farming enhances the agriculture system as a whole. It maintains a number of variables, including soil, temperature, and humidity. The application of ML in the agricultural sector aims to raise crop quality and productivity. Appropriate algorithmic application senses data and aids in crop selection. Comparing different ML algorithms aids to improve the system accuracy. Thus, the technology is used to lessen the challenges farmers encounter and raise the amount and caliber of labor they produce [21].

Bhoi et al. [22] show IoT- and ML-based irrigation system. Regression and classification models based on ML are integrated with intelligence. The weather forecast data utilizing APIs further strengthen the robustness of our suggested solution. Based on all the experiments, it is determined that the K-Nearest Neighbors (KNN) classifier trained system in conjunction with the suggested support vector regression (SVR) is incredibly effective for this difficult task.

IoT-based smart city applications demand increased data output and processing [23]. The suggested approach based on data gathered from IoT sensor devices, DBN-R-LSTM-NN, is employed for fire outbreak categorization and prediction. The Arduino MCU serves as the brain for collecting data from the vast array of IoT devices, whereas the data preprocessor analyzes the features for classification. Temperature, Flame, Smoke, Fine Fuel Moisture Code (FFMC), Relative Humidity (RH), and Duff Moisture Code (DMC) are the primary characteristics [24].

Malki et al. [25] address urban air quality challenges by predicting the Air Quality Index (AQI) using ML Algorithms: Neural Networks and Support Vector Machines. Utilizing data from the Central Pollution Control Board, the proposed model demonstrates promising results in predicting Delhi's AQI, showcasing enhanced accuracy. The approach holds potential for application in other smart cities, contributing to pollution mitigation and public health improvement [26].

Brahim [27] investigates the impact of weather variables on COVID-19 transmission using regressor ML models. Analyzing temperature and humidity, the models reveal their relevance in predicting the mortality rate compared to other census variables. Experimental results demonstrate that higher temperatures correlate with lower infection cases, emphasizing the significance of weather conditions in COVID-19 outcomes. Machorro-Cano et al. [28] address the global issue of AP and emphasizes the significance of real-time monitoring using IoT sensors. The integration of ML algorithms enhances AP prediction accuracy, offering a cost-effective and dynamic solution. The study reviews various ML techniques applied to AP prediction, discussing their merits, demerits, and challenges in real-time monitoring across different cities.

An ML approach to forecast electricity consumption based on weather data was introduced [29–35], offering a potential solution for households and smart cities to manage electricity usage effectively. Utilizing a dataset with appliance readings and weather conditions, the method achieves a 75.7% correlation coefficient using Random Tree. The results pave the way for the development of an accurate smart electricity consumption monitor, aiding households in optimizing electricity usage by considering weather conditions.

HEMS-IoT, a smart home energy management system utilizing big data and ML for energy efficiency, was introduced [36–38]. Focused on home comfort, safety, and energy savings, the system employs the ML algorithm to analyze user behaviors and energy consumption patterns. RuleML and Apache Mahout are utilized to generate energy-saving recommendations based on user preferences, ensuring smart home optimization. The case study demonstrates the system's effectiveness in monitoring a smart home, validating its potential for enhancing comfort, safety, and reducing energy consumption.

3.2 CASE STUDY: WEATHER MONITORING AND FORECASTING SYSTEM

This chapter proposes an IoT-based weather monitoring and forecasting system tailored specifically for smart cities. Figure 3.1 depicts an outlook of the device, demonstrating how the sensors are utilized at a given location to collect information on the weather characteristics. The focus is on integrating ML algorithms into the system to bolster forecasting accuracy and monitoring capabilities. These algorithms are meticulously selected and tailored to suit the dynamic and complex nature of weather patterns in urban environments.

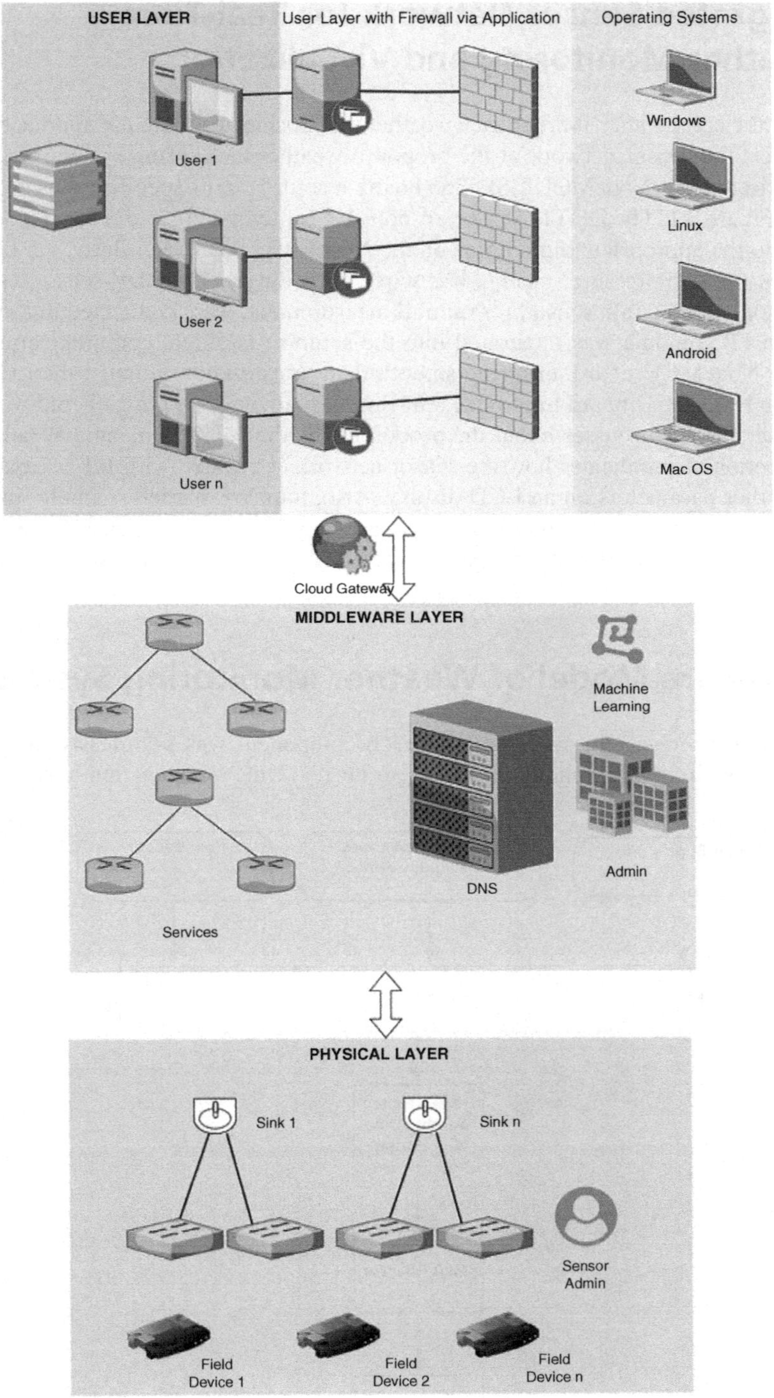

FIGURE 3.1 Overview of the weather monitoring system.

3.2.1 Integrated Sensor Network for Real-Time Weather Monitoring and Visualization

To interconnect the components for real-time weather monitoring, a systematic approach was followed. Figure 3.2 depicts the sensor network of the proposed weather monitoring system. First, each sensor was carefully wired to the NodeMCU ESP8266 board according to its specifications and requirements as indicated in Figure 3.1. The DHT11 sensor, responsible for temperature and humidity measurement, was connected to the appropriate digital pins on the NodeMCU board. Similarly, the BMP180 sensor, dedicated to atmospheric pressure sensing, was wired accordingly. The LDR sensor, utilized for light intensity detection, and the rain sensor, for rainfall measurement, were connected in a similar manner. Additionally, an I2C module was integrated into the setup to facilitate communication between the sensors and the NodeMCU board, ensuring smooth data transmission. Finally, the LCD display was connected to the NodeMCU board to provide real-time visualization of the gathered weather data.

This chapter further discusses about the proposed IoT and ML Augmented Weather Monitoring System. This section also indicates how the sensor network associated with IoT is capable of displaying the key weather parameters on an LCD display. Also, four ML models, namely linear regression, decision tree, random forest, and lasso regression, are deployed on the dataset for weather forecasting. Error Metrics and the scatter plot suggest the best ML model for the considered dataset spanning around 10 years.

3.2.2 Hardware Model of Weather Monitoring System

Through meticulous wiring and configuration, each component was seamlessly integrated into the hardware setup, enabling comprehensive and accurate real-time weather monitoring capabilities as

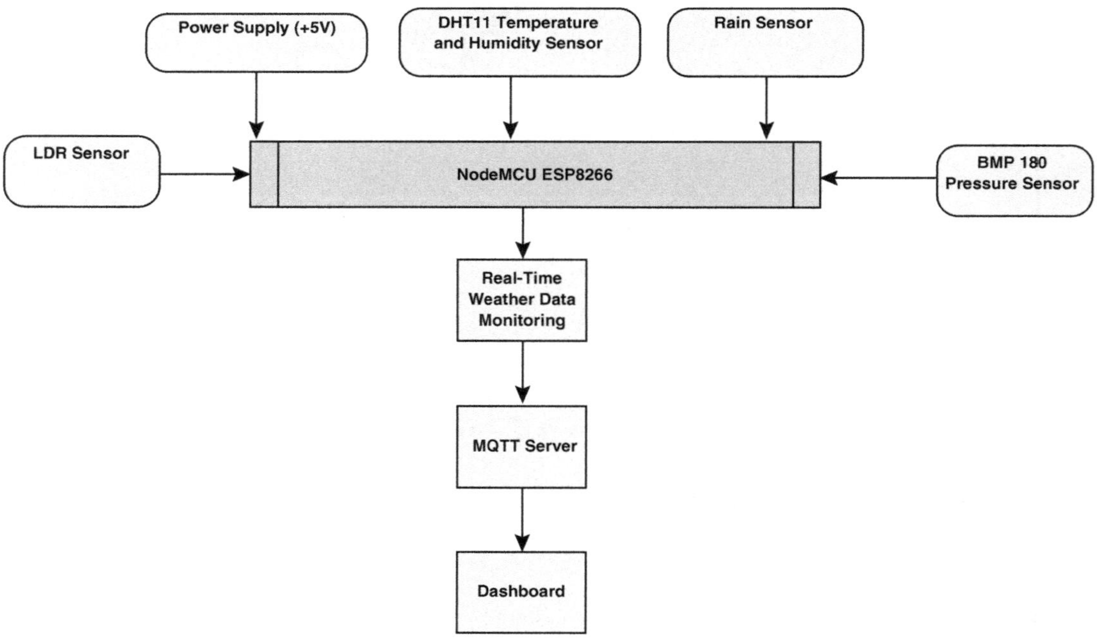

FIGURE 3.2 Sensor network for weather monitoring system.

shown in Figure 3.3. Configure WiFi and MQTT settings initialize connections, set up callback functions for message handling, and display data on the LCD for real-time monitoring on the Adafruit dashboard as shown in Table 3.1.

3.2.3 Weather Forecasting Using Machine Learning Algorithms

The dataset employed is from Kaggle featuring "Historical Weather Data for Indian Cities." Specifically focusing on Hyderabad City in India, we curated this dataset to address the community's demand for comprehensive historical weather records. Selected datasets encompass the top eight Indian cities by population, spanning from January 1, 2009, to January 1, 2022, providing over a decade's worth of hourly weather data as shown in Table 3.2. Extraction utilized the worldweatheronline.com API and the wwo_hist package. This extensive repository facilitates diverse applications, from visualizing the impacts of climate change to forecasting weather patterns for upcoming days, weeks, months, and seasons.

FIGURE 3.3 Hardware model of weather monitoring system

TABLE 3.1 Real-time display of sensor values on dashboard by Adafruit IO

DATE AND TIME	*PARAMETER*	*VALUE*
2024-04-10 12:21:00	Light	5.00
2024-04-10 12:21:00	Humidity	80.00
2024-04-10 12:21:00	Temperature	29.90
2024-04-10 12:21:00	Pressure	1004.00
2024-04-10 12:21:00	Rain	Rainfall is absent

TABLE 3.2 Historical weather dataset of Hyderabad City, India

DATE_TIME														
2009-01-01 00:00:00	28	15	0.0	8.7	6	1	31	09.:57 AM	09:58 PM	06:46 AM	---	21	9
2009-01-01 01:00:00	28	15	0.0	8.7	6	1	31	09.:57 AM	09:58 PM	06:46 AM	---	20	9
2009-01-01 02:00:00	28	15	0.0	8.7	6	1	31	09.:57 AM	09:58 PM	06:46 AM	---	20	8
2009-01-01 03:00:00	28	15	0.0	8.7	6	1	31	09.:57 AM	09:58 PM	06:46 AM	---	19	8
2009-01-01 04:00:00	28	15	0.0	8.7	6	1	31	09.:57 AM	09:58 PM	06:46 AM	---	21	7

3.2.3.1 Datasets

Dataset have various number of parameters measured over 10 years and recorded data for every one hour. But we are considering few of the parameters for training the data. Those are

```
Index['maxtempC', 'mintempC', 'totalSnow_cm', 'sunHour', 'uvIndex',
'uvIndex.1', 'moon_illumination', 'moonrise', 'moonset', 'sunrise',
'sunset', 'DewPointC', 'FeelsLikeC', 'HeatIndexC', 'WindChillC',
'WindGustKmph', 'cloudcover', 'humidity', 'precipMM', 'pressure',
'tempC', 'visibility', 'winddirDegree', 'windspeedKmph']
```

Figure 3.4 shows how the model is trained with the help of histograms and plots.

3.2.3.2 Error analysis for weather forecasting using ML algorithms

In conducting error analysis for weather prediction using ML algorithms, the dataset is divided into training (80%) and testing sets (20%). ML models such as linear regression, decision tree, random forest, or lasso regression are trained using the training data to learn patterns between input features like humidity and target variables like temperature. Predictions are then made on the testing dataset whose results for the four ML models are as shown in Tables 3.3–3.6. The error metrics such as Mean Absolute Error (MAE), MSE, and R-squared (R^2) are calculated to assess model performance as shown in Table 3.7. Lower MAE and MSE values indicate better accuracy, while higher R^2 values signify better fit. Comparing error metrics helps identify the best-performing model, aiding in informed decision-making for weather prediction. Iterative adjustments may be made to improve model performance until satisfactory results are achieved.

By examining Figure 3.5, the random forest algorithm is identified best for predicting values from the dataset. This choice is underpinned by its remarkable 98% accuracy rate, the MAE of 0.39, residual sum of squares (MSE) of 0.50, and R^2-score of 0.98, which indicate high accuracy in the forecasting model surpassing other algorithms in performance.

3.3 CONCLUSION

A case study of Hyderabad, India, is considered for the weather forecasting systems. With the successful completion of the hardware setup, the IoT- and ML-aided weather monitoring system adeptly displays a comprehensive array of weather parameters including temperature, pressure, light intensity, humidity, and rainfall on both the LCD display and the Adafruit dashboard. After deploying four

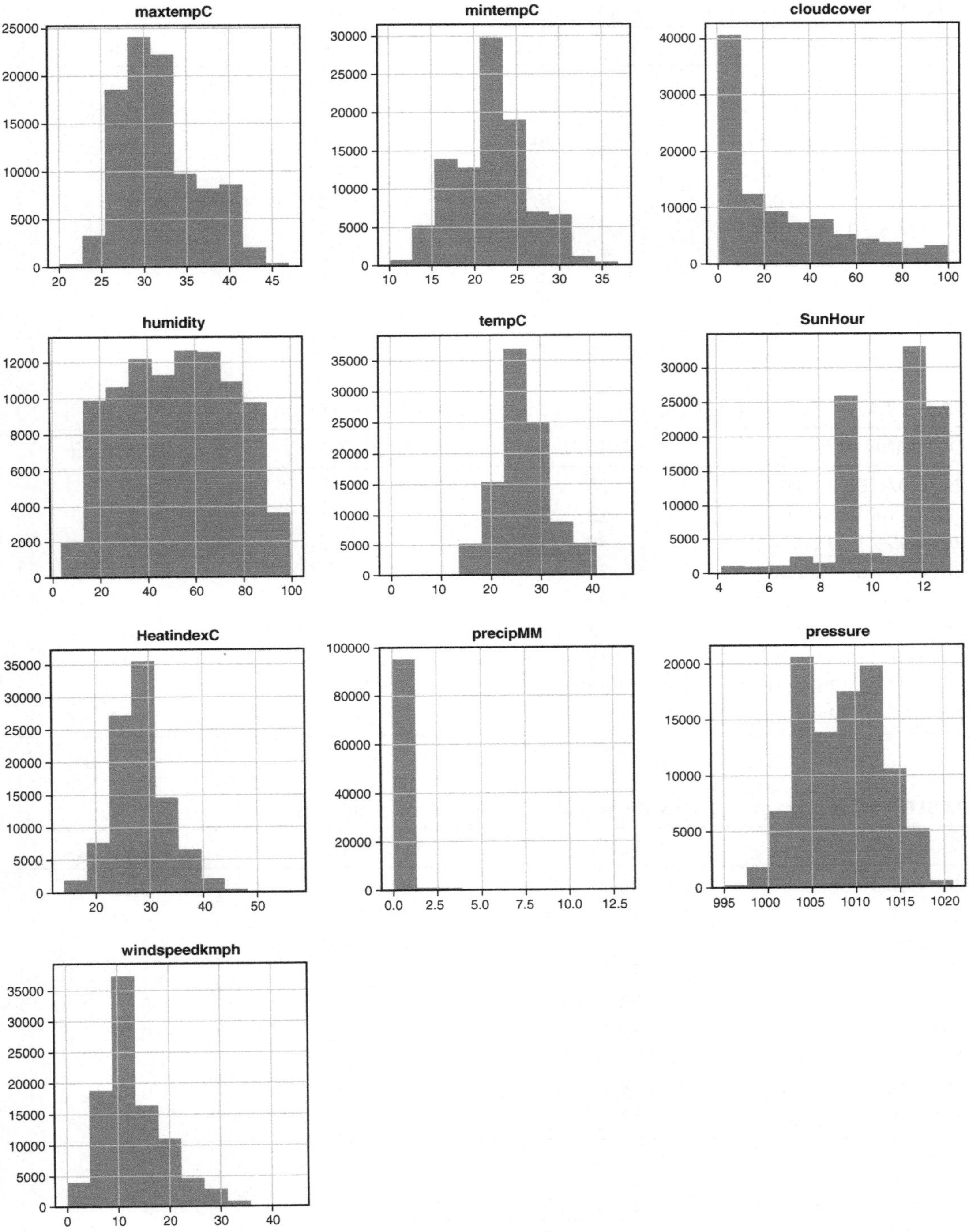

FIGURE 3.4 Training the model with the help of histograms and plots.

TABLE 3.3 Performance of linear regression on temperature parameter

DATE AND TIME	*ACTUAL*	*PREDICTED*	*DIFFERENCE*
2024-07-10 08:00:00	25	25.01	–0.01
2024-11-04 20:00:00	23	23.17	–0.17
2024-09-21 09:00:00	29	28.93	0.07
2024-02-16 11:00:00	30	28.12	1.88
2024-07-21 01:00:00	23	22.82	0.18
2023-03-30 09:00:00	33	32.93	0.07
2019-12-31 05:00:00	18	16.95	1.05
2019-08-02 17:00:00	23	22.84	0.16
2015-11-12 12:00:00	30	29.09	0.91

TABLE 3.4 Performance of linear regression on decision tree regression

DATE AND TIME	*ACTUAL*	*PREDICTED*	*DIFFERENCE*
2024-07-10 08:00:00	25	25	0
2024-11-04 20:00:00	23	23	0
2024-09-21 09:00:00	29	29	0
2024-02-16 11:00:00	30	30	0
2024-07-21 01:00:00	23	23	0
2023-03-30 09:00:00	33	33	0
2019-12-31 05:00:00	18	18	0
2019-08-02 17:00:00	23	23	0
2015-11-12 12:00:00	30	28	2.0

TABLE 3.5 Performance of linear regression on random forest regression

DATE AND TIME	*ACTUAL*	*PREDICTED*	*DIFFERENCE*
2024-07-10 08:00:00	25	24.94	0.06
2024-11-04 20:00:00	23	22.57	0.43
2024-09-21 09:00:00	29	29	0
2024-02-16 11:00:00	30	30.18	–0.18
2024-07-21 01:00:00	23	23	0
2023-03-30 09:00:00	33	33.18	–0.18
2019-12-31 05:00:00	18	18	0
2019-08-02 17:00:00	23	22.92	0.08
2015-11-12 12:00:00	30	30.19	–0.19

TABLE 3.6 Performance of linear regression on lasso regression

DATE AND TIME	*ACTUAL*	*PREDICTED*	*DIFFERENCE*
2024-07-10 08:00:00	25	24.99	0.01
2024-11-04 20:00:00	23	23.12	–0.12

(Continued)

TABLE 3.6 (*Continued*) Performance of linear regression on lasso regression

DATE AND TIME	*ACTUAL*	*PREDICTED*	*DIFFERENCE*
2024-09-21 09:00:00	29	28.73	0.27
2024-02-16 11:00:00	30	30.18	–0.18
2024-07-21 01:00:00	23	23	0
2023-03-30 09:00:00	33	28.17	1.83
2019-12-31 05:00:00	18	16.92	1.08
2019-08-02 17:00:00	23	23.27	–0.27
2015-11-12 12:00:00	24	22.34	1.66

TABLE 3.7 Error metrics

ML MODEL	*MEAN ABSOLUTE ERROR*	*RESIDUAL SUM OF SQUARES*	*R^2-SCORE*
Linear regression	0.85	1.46	0.95
Decision tree regression	0.43	0.87	0.97
Random forest algorithm	0.39	0.50	0.98
Lasso regression	0.86	1.56	0.95

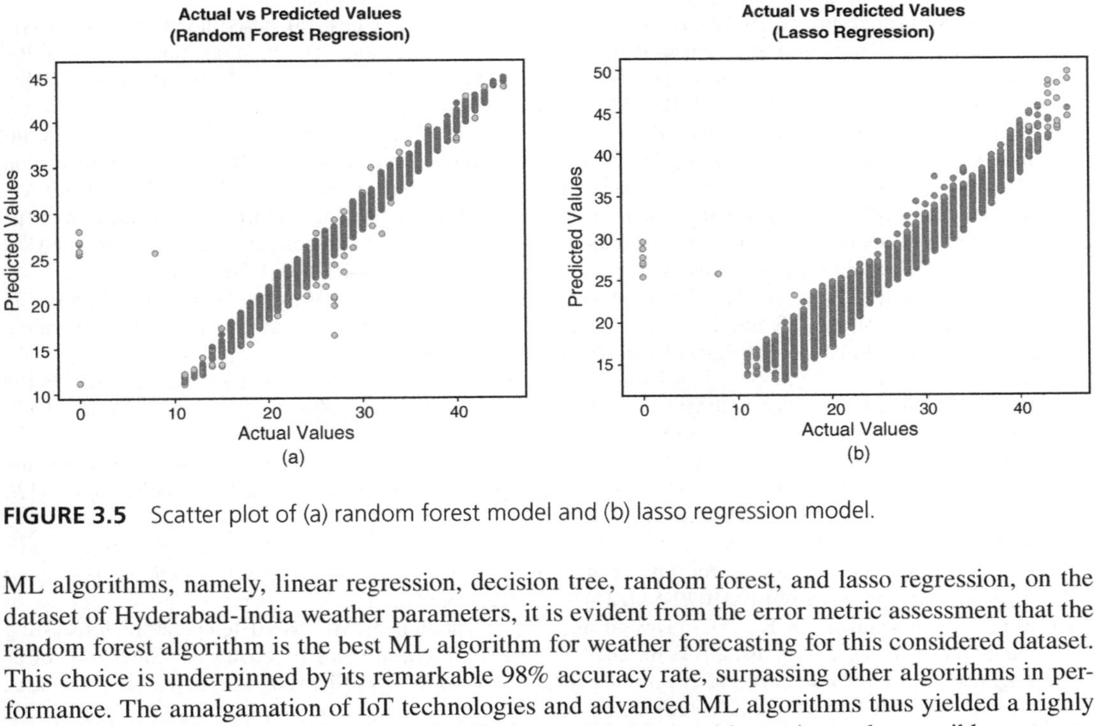

FIGURE 3.5 Scatter plot of (a) random forest model and (b) lasso regression model.

ML algorithms, namely, linear regression, decision tree, random forest, and lasso regression, on the dataset of Hyderabad-India weather parameters, it is evident from the error metric assessment that the random forest algorithm is the best ML algorithm for weather forecasting for this considered dataset. This choice is underpinned by its remarkable 98% accuracy rate, surpassing other algorithms in performance. The amalgamation of IoT technologies and advanced ML algorithms thus yielded a highly promising weather monitoring system, furnishing stakeholders with precise and accessible meteorological forecasts and invaluable insights.

As a future scope, the proposed IoT- and ML-aided weather monitoring and forecasting system finds key applications: real-time alerts, disaster management, smart agriculture, marine safety, environmental monitoring, stand-alone renewable energy systems, automation, and predictive analytics.

REFERENCES

1. Joseph, F.J., 2019. IoT based weather monitoring system for effective analytics. *International Journal of Engineering and Advanced Technology*, 8(4), pp. 311–315.
2. Patel, A. and Swathika, O.G., 2024. Analyzing the predictive power of solar photovoltaics using three machine learning models. In: *2024 3rd Odisha International Conference on Electrical Power Engineering, Communication and Computing Technology (ODICON)*, Bhubaneswar, India, pp. 1–6. IEEE. doi:10.1109/ODICON62106.2024.10797527.
3. Girija, C., Grace, S.A., Harshalatha, H. and Pushpalatha, H.P., 2018. Internet of things (IoT) based weather monitoring system. *International Journal of Engineering Research & Technology*, 6(13), p. 4. doi:10.17577/IJERTCONV6IS13149.
4. Bhagat, A.M., Thakare, A.G., Molke, K.A., Muneshwar, N.S. and Choudhary, P.V., 2019. IoT based weather monitoring and reporting system project. *International Journal of Trend in Scientific Research and Development*, 3(3), pp. 1–3. doi:10.31142/ijtsrd21677.
5. Patel, A. and Swathika, O.G., 2024. Off-grid small-scale power forecasting using optimized machine learning algorithms. *IEEE Access*, 12. doi:10.1109/ACCESS.2024.3430385.
6. Patel, A. and Gnana Swathika, O.V., 2023. Day-ahead solar power forecasting using statistical and machine learning methods. *Integrated Green Energy Solutions,* 2, pp. 71–101. doi:10.1002/9781394193738.ch26.
7. Meera, S., Sharmikha Sree, R.B., Kalpana, R.A., Manasvinii, S.R., Haritha, V. and Valarmathi, K., 2021. IoT based weather reporting system using Arduino and node MCU. *Smart Intelligent Computing and Communication Technology*, 38, p. 209. doi:10.3233/APC210038.
8. Raju, M.P. and Laxmi, A.J., 2020. IoT based online load forecasting using machine learning algorithms. *Procedia Computer Science*, 171, pp. 551–560. doi:10.1016/j.procs.2020.04.059.
9. Rani, D.S., Jayalakshmi, G.N. and Baligar, V.P., 2020, March. Low cost IoT based flood monitoring system using machine learning and neural networks: flood alerting and rainfall prediction. In: *2020 2nd International Conference on Innovative Mechanisms for Industry Applications (ICIMIA)*, Bangalore, India, pp. 261–267. IEEE. doi:10.1109/ICIMIA48430.2020.9074928.
10. Dahane, A., Benameur, R., Kechar, B. and Benyamina, A., 2020, October. An IoT based smart farming system using machine learning. In: *2020 International Symposium on Networks, Computers and Communications (ISNCC)*, Montreal, Canada, pp. 1–6. IEEE. doi:10.1109/ISNCC49221.2020.9297341.
11. Majumdar, P., Mitra, S. and Bhattacharya, D., 2021. IoT for promoting agriculture 4.0: a review from the perspective of weather monitoring, yield prediction, security of WSN protocols, and hardware cost analysis. *Journal of Biosystems Engineering*, 46(4), pp. 440–461. doi:10.1007/s42853-021-00118-6.
12. Price, I., Sanchez-Gonzalez, A., Alet, F., Andersson, T.R., El-Kadi, A., Masters, D., Ewalds, T., Stott, J., Mohamed, S., Battaglia, P. and Lam, R., 2025. Probabilistic weather forecasting with machine learning. *Nature*, 637(8044), pp. 84–90. doi:10.1038/s41586-024-08252-9.
13. Sankaranarayanan, S., Prabhakar, M., Satish, S., Jain, P., Ramprasad, A. and Krishnan, A., 2020. Flood prediction based on weather parameters using deep learning. *Journal of Water and Climate Change*, 11(4), pp. 1766–1783. doi:10.2166/wcc.2019.321.
14. Maduranga, M.W.P. and Abeysekera, R., 2020. Machine learning applications in IoT based agriculture and smart farming: a review. *International Journal of Engineering Applied Sciences and Technology*, 4(12), pp. 24–27. 10.33564/IJEAST.2020.v04i12.004.
15. Onesimu, J.A., Kadam, A., Sagayam, K.M. and Elngar, A.A., 2021. Internet of things based intelligent accident avoidance system for adverse weather and road conditions. *Journal of Reliable Intelligent Environments*, pp. 1–15. doi:10.1109/I-SMAC47947.2019.9032662.
16. Borah, B. and Sarma, K.K., 2021. Design and implementation of ML based temperature forecasting model for air conditioning using IoT. In: *2021 1st Odisha International Conference on Electrical Power Engineering, Communication and Computing Technology (ODICON)*, Bhubaneswar, India, pp. 1–6. IEEE. doi:10.1109/ODICON50556.2021.9428934.
17. Kanade, P. and Prasad, J.P., 2021. Arduino based machine learning and IoT smart irrigation system. *International Journal of Soft Computing and Engineering*, 10(4), pp. 1–5. doi:10.35940/ijsce.D3481.0310421.
18. Wayangkau, I.H., Mekiuw, Y., Rachmat, R., Suwarjono, S. and Hariyanto, H., 2020. Utilization of IoT for soil moisture and temperature monitoring system for onion growth. *Emerging Science Journal*, 4, pp. 102–115. 10.28991/esj-2021-SP1-07.

19. Tejaswi, P. and Swathika, O.G., 2023. Machine learning algorithms-based solar power forecasting in smart cities. In: S. Dey & A. E. Hassanien (Eds.), *Artificial Intelligence and Machine Learning in Smart City Planning*, pp. 171–179, Elsevier. doi:10.1016/B978-0-323-99503-0.00001-6.
20. Podder, A.K., Al Bukhari, A., Islam, S., Mia, S., Mohammed, M.A., Kumar, N.M., Cengiz, K. and Abdulkareem, K.H., 2021. IoT based smart agrotech system for verification of Urban farming parameters. *Microprocessors and Microsystems*, 82. doi:10.1016/j.micpro.2021.104025.
21. Swathika, O.G. and Hemapala, K.T.M.U., 2019. IoT based energy management system for stand-alone PV systems. *Journal of Electrical Engineering & Technology*, 14(5), pp. 1811–1821. doi:10.1007/s42835-019-00193-y.
22. Bhoi, A., Nayak, R.P., Bhoi, S.K. and Sethi, S., 2021. Automated precision irrigation system using machine learning and IoT. In: A. E. Hassanien et al. (Eds.), *Intelligent Systems: Proceedings of ICMIB 2020*, pp. 275–282, Springer. doi:10.1007/978-981-33-6081-5_24.
23. Zhang, Y., Geng, P., Sivaparthipan, C.B. and Muthu, B.A., 2021. Big data and artificial intelligence based early risk warning system of fire hazard for smart cities. *Sustainable Energy Technologies and Assessments*, 45. doi:10.1016/j.seta.2020.100986.
24. Lee, M., Lin, L., Chen, C.Y., Tsao, Y., Yao, T.H., Fei, M.H. and Fang, S.H., 2020. Forecasting air quality in Taiwan by using machine learning. *Scientific Reports*, 10(1), p. 4153.
25. Malki, Z., Atlam, E.S., Hassanien, A.E., Dagnew, G., Elhosseini, M.A. and Gad, I., 2020. Association between weather data and COVID-19 pandemic predicting mortality rate: machine learning approaches. *Chaos, Solitons & Fractals*, 138. doi:10.1016/j.chaos.2020.110137.
26. Kalaivani, G. and Mayilvahanan, P., 2021. Air quality prediction and monitoring using machine learning algorithm based IoT sensor-A researcher's perspective. In: *2021 6th International Conference on Communication and Electronics Systems (ICCES)*, Coimbatore, India, pp. 1–9. IEEE. doi:10.1109/ICCES51350.2021.9489153.
27. Brahim, G.B., 2021. Weather conditions impact on electricity consumption in smart homes: machine learning based prediction model. In: *2021 8th International Conference on Electrical and Electronics Engineering (ICEEE)*, Antalya, Turkey, pp. 93–98. IEEE. doi:10.1109/ICEEE52452.2021.9415917.
28. Machorro-Cano, I., Alor-Hernández, G., Paredes-Valverde, M.A., Rodríguez-Mazahua, L., Sánchez-Cervantes, J.L. and Olmedo-Aguirre, J.O., 2020. HEMS-IoT: a big data and machine learning-based smart home system for energy saving. *Energies*, 13(5), p. 1097. doi:10.3390/en13051097.
29. Bochenek, B. and Ustrnul, Z., 2022. Machine learning in weather prediction and climate analyses—applications and perspectives. *Atmosphere*, 13(2), p. 180. doi:10.3390/atmos13020180.
30. Chen, L., Han, B., Wang, X., Zhao, J., Yang, W. and Yang, Z., 2023. Machine learning methods in weather and climate applications: a survey. *Applied Sciences*, 13(21). doi:10.20944/preprints202309.1764.v2.
31. Najian, M. and Goudarzi, N., 2023. Evaluating critical weather parameters using machine learning models. *ASME Power Conference*, 87172, V001T04A006. doi:10.1115/POWER2023-108893.
32. Kumari, S. and Muthulakshmi, P., 2023. A wide scale survey on weather prediction using machine learning techniques. *Journal of Information & Knowledge Management*, 22(5). doi:10.1142/S0219649222500939.
33. Markovics, D. and Mayer, M.J., 2022. Comparison of machine learning methods for photovoltaic power forecasting based on numerical weather prediction. *Renewable and Sustainable Energy Reviews*, 161. https://doi.org/10.1016/j.rser.2022.112364.
34. Chandrayan, S.S., Khushal, S., and Akash Kumar, B. 2022. Atmospheric weather fluctuation prediction using machine learning. In: A. E. Hassanien et al. (Eds.), *Cognitive Informatics and Soft Computing: Proceeding of CISC 2021*, pp. 431–443, Springer Nature.
35. Choi, S. and Jung, E.S., 2023. Optimizing numerical weather prediction model performance using machine learning techniques. *IEEE Access*, 11. doi:10.1109/ACCESS.2023.3297200.
36. Scher, S., & Messori, G. (2018). Predicting weather forecast uncertainty with machine learning. *Quarterly Journal of the Royal Meteorological Society*, 144(717), 2830–2841. doi:10.1002/qj.3410.
37. Wang, B., Lu, J., Yan, Z., Luo, H., Li, T., Zheng, Y. and Zhang, G., 2019, July. Deep uncertainty quantification: a machine learning approach for weather forecasting. In: *Proceedings of the 25th ACM SIGKDD International Conference on Knowledge Discovery & Data Mining*, Anchorage, Alaska, pp. 2087–2095. doi:10.1145/3292500.3330704.
38. Siddique, T., Mahmud, M.S., Keesee, A.M., Ngwira, C.M. and Connor, H., 2022. A survey of uncertainty quantification in machine learning for space weather prediction. *Geosciences*, 12(1), p. 27. doi:10.3390/geosciences12010027.

Toward Industry 4.0
PLC Automation for Industrial Applications

4

G. Surya Deepan, G. Jagadish, R. Guru Karthik, and O. V. Gnana Swathika

4.1 INTRODUCTION

Programmable Logic Controllers (PLCs) are an integral part of modern industrial systems, providing unparalleled reliability, flexibility, and efficiency in controlling complex processes. Widely adopted across various industries, PLCs are renowned for their ability to automate tasks, optimize performance, and ensure safety in dynamic and demanding environments. Their versatility makes them a cornerstone of technological advancements in Industrial Automation, Control Systems, Smart Grids, Smart Cities, and Robotics. By streamlining processes and reducing manual intervention, PLCs significantly improved productivity and operational efficiency, which in turn transforms traditional systems into more intelligent and adaptive frameworks. This chapter explores the extensive applications of PLC-Rockwell ControlLogix 5580 in these domains, shedding light on their transformative impact. It delves into the unique capabilities of PLCs, such as real-time monitoring, precise control, and seamless integration with advanced technologies. A key highlight of this work is the proposal of a novel PLC-based temperature and pressure control system that incorporates industrial-grade temperature, gas, and pressure sensors. This system demonstrates the precision and adaptability of PLCs in optimizing industrial operations. Designed to enhance the performance of Heating, Ventilation, and Air Conditioning (HVAC) systems and pressure ducts, the proposed system ensures consistent and reliable operation under varying industrial conditions, reducing energy consumption while maintaining operational stability. To realize this concept, the system leverages real-time data from the sensors to regulate temperature and pressure in demanding industrial environments, effectively addressing the challenges of dynamic load variations and environmental fluctuations. Further, this chapter also provides insights into the future of PLC technologies within the framework of Industry 4.0. As industries increasingly embrace digitalization, PLCs are evolving to integrate with emerging technologies such as Internet of Things (IoT), AI, and cloud computing, enabling more robust

DOI: 10.1201/9781003649700-4

automation and interoperability. Addressing challenges such as cybersecurity and processing power will further enhance PLC capabilities, cementing their role as drivers of innovation in the industrial landscape.

4.1.1 Literature Review

The recent papers explore innovative applications of PLCs across diverse domains, emphasizing their importance in automation and control. A novel approach to photovoltaic maximum power point tracking (MPPT) using robust sliding mode control, implemented via Siemens S7-1200 PLC, demonstrates smoother and more precise performance compared to the incremental conductance algorithm, particularly under varying meteorological conditions [1]. Another study presents a water-level control system for solar water heating projects using PLCs, industrial touchscreens, and variable-frequency drives. This system enhances scalability, remote monitoring, and energy efficiency, which offers significant improvements over traditional controllers [2]. A smart traffic control system designed with Siemens PLC SIMATIC S7-300 and SCADA optimizes traffic flow using motion sensors and adaptive light durations, effectively reducing congestion and improving efficiency compared to conventional systems [3]. Lastly, a smart streetlighting system integrates PLC technology in wired and hybrid communication modes, supporting adaptive dimming, autonomous control, and advanced connectivity. This system underscores the potential of smart streetlights as a foundation for smart city applications while addressing cybersecurity vulnerabilities [4]. A Mitsubishi Melsec PLC-based system automates multi-machine operations for sorting and packaging products by color, integrating sensors, DC motors and solenoids. This setup enhances efficiency and reduces human error, with real-time monitoring through Human-Machine Interface (HMI) [5]. Another study explores optimal pulse frequency and settings for stepper motor control using PLCs, concluding that a frequency of 5000 Hz and 100,000 pulses offer precision and efficiency, making PLCs suitable for various industrial applications [6]. An anomaly detection methodology employing a one-class support vector machine (OCSVM) focuses on optimizing model parameters for PLC forensic investigations. The study emphasizes the importance of precise timestamps and historian data for improving detection accuracy [7]. Lastly, supervisory control for flexible manufacturing systems (FMS) is implemented using the theory of regions and Petri nets, showcasing PLCs' capability in managing asynchronous, event-driven systems. The study proposes a control policy for improved performance and usability while recommending interface enhancements and direct PLC code visualization to streamline industrial automation [8]. One paper automates the movement of a platform using a Gyro sensor and PLC, offering precise angle adjustments with potential IoT integration for enhanced efficiency [9]. Another develops a Proportional Integral-Derivative (PID) control system for biodiesel reactors, effectively managing temperature and motor speed to improve biodiesel production quality using an OMRON PLC [10]. A low-cost protection scheme for three-phase induction motors uses PLCs and HMI to monitor parameters like overcurrent and overvoltage, replacing traditional relays with solid-state ones for improved reliability and real-time fault protection [11]. Lastly, a temperature control system for electric furnaces integrates digital PID algorithms with a Mitsubishi PLC and a touch screen for real-time monitoring and user-friendly operation, enhancing accuracy and reducing energy waste [12].

4.2 PROPOSED ARCHITECTURE OF THE PLC CONTROL SYSTEM:

This chapter explores the use of PLC-Rockwell ControlLogix 5580 and sensors to automate temperature and pressure control in industrial and commercial buildings. By integrating sensors with PLCs, real-time data on temperature and pressure rates are monitored, allowing precise regulation of HVAC systems and pressure distribution in high-pressure chambers. The system improves energy efficiency, reduces operational costs, and enhances indoor environmental quality. Additionally, the scalability of PLCs allows easy adaptation to varying building sizes and complexities. This solution addresses the growing need for intelligent energy management, contributing to sustainability and improved performance in building automation systems.

4.2.1 Problem Statement

4.2.1.1 Topic

This research addresses the problem of poor temperature and pressure management in industrial and commercial buildings.

4.2.1.2 Context

Although recent studies have explored various methods in optimizing energy using temperature and pressure control schemes, incorporation of PLCs and sensors is rarely used despite their efficiency and flexibility.

4.2.1.3 Knowledge gap

Current methods focus on small-scale smart houses and smart rooms. A solution for a larger scale and volume is yet to be explored.

4.2.1.4 Problem definition

This study investigates the impact of hybrid usage of technologies such as PLCs, HMIs, sensors, and HVAC concepts to optimize temperature and power flow among commercial-scale buildings.

4.2.1.5 Impact

Solving this issue will reduce energy usage and cost, which aids the consumer and user to be more comfortable. It could also be used to increase efficiency of machineries in manufacturing units.

4.2.2 Block Diagram

Figure 4.1 shows the process flows from environment monitoring through sensors, signal conversion (analog-to-digital converter (ADC) and digital-to-analog converter (DAC)), decision-making (PLC), and interaction (HMI), culminating in action through output devices to control temperature and pressure. This loop ensures continuous, automated monitoring and control to achieve desired environmental conditions in industrial settings.

4.2.2.1 Environment

This block represents the industrial setting where temperature and pressure need to be controlled. The environment includes all external factors that affect temperature and pressure, such as machinery heat, air quality, and external climate conditions. Monitoring these conditions is essential as fluctuations can impact productivity, safety, and equipment life.

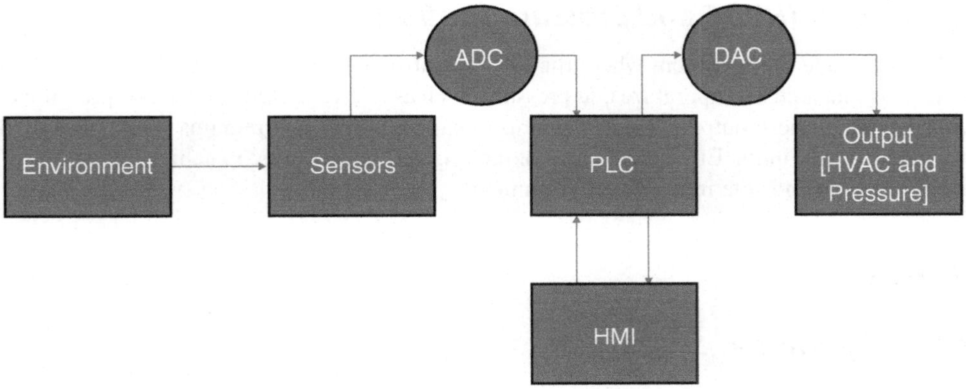

FIGURE 4.1 System architecture of the case study.

4.2.2.2 Sensors

Sensors are responsible for collecting real-time data on temperature, pressure, and possibly humidity from the environment. Typical sensors include thermocouples or Resistance Temperature Detector (RTDs) (for temperature), pressure transducers, and humidity sensors. The raw data collected by these sensors is analog in nature and needs to be converted into digital form for processing by the PLC.

4.2.2.3 Analog-to-digital converter

The ADC converts the analog signals from the sensors into digital signals that the PLC processes. This conversion is crucial because PLCs operate digitally, and analog signals must be digitized to be interpreted accurately. The quality and accuracy of the ADC significantly impact the overall control precision, especially in high-sensitivity environments.

4.2.2.4 Programmable logic controller

The PLC is the core processing unit of the system responsible for making decisions based on sensor data. The digital data from the ADC is input to the PLC, where it is compared to predefined setpoints for temperature and pressure. Based on these comparisons, the PLC executes control actions, adjusting outputs to maintain desired conditions. The PLC is programmed with various control algorithms, such as PID control to optimize performance. It sends control signals to the actuators and interfaces with the DAC and HMI.

4.2.2.5 Human-machine interface

The HMI allows operators to monitor and interact with the control system. It displays real-time data from sensors, including current temperature, pressure, and system status. Operators use the HMI to adjust setpoints, view alerts, and troubleshoot issues. The HMI improves user accessibility and provides a way for personnel to make adjustments to the system based on changing operational needs.

4.2.2.6 Digital-to-analog converter

After the PLC makes a decision, the DAC converts digital control signals back into analog signals if required by the output devices. For instance, certain actuators that regulate HVAC systems may require analog inputs. The DAC plays a vital role in ensuring that control signals are compatible with output devices.

4.2.2.7 Output (HVAC and pressure control)

This is the final stage of the system where the PLC's control actions take effect. Outputs may include various devices that adjust temperature and pressure, such as HVAC systems, fans, and pressure control valves. By adjusting these outputs, the system ensures the environment remains within specified temperature and pressure limits. Effective control of these outputs is essential to achieving the objective of maintaining a stable and safe industrial environment.

4.2.3 Algorithm

4.2.3.1 Temperature

Figure 4.2 illustrates an HVAC control process with several steps to monitor, adjust, and maintain optimal environmental conditions.

1. **Start and Initialization**: The process begins with a "Start" command, followed by "Power ON," where the system activates.
2. **Environment Scanning and Temperature Decision**: The system scans the surrounding environment to assess the current conditions and subsequently decides the optimal temperature settings.
3. **Activation of HVAC Control**: The system then turns ON HVAC control to adjust the temperature based on predefined scaling parameters.
4. **Monitoring and Adjusting**: Once HVAC control is active, the system continuously monitors and, if necessary, adjusts the temperature. If the temperature setting process is complete, it turns OFF the HVAC control.

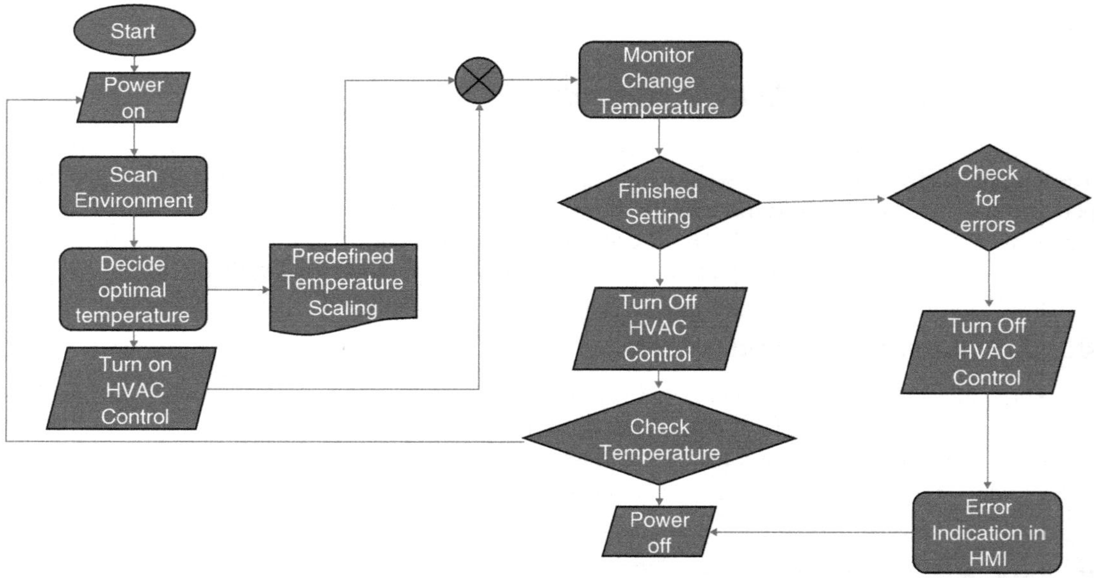

FIGURE 4.2 Algorithm for temperature control.

5. **Error Checking**: The system checks for any potential errors. If no errors are detected, it continues to monitor the temperature. Otherwise, the system will turn OFF the HVAC control and provide an error indication on the HMI.
6. **Shutdown**: After temperature monitoring is completed without issues, the system proceeds to power OFF mode.

This process ensures that temperature regulation is achieved efficiently with an integrated error-checking mechanism to alert operators if problems arise. The flowchart is likely intended for industrial or building automation settings where HVAC control needs to be precise and reliable.

4.2.3.2 Pressure

Figure 4.3 outlines a pressure control process using pressure ducts, designed to monitor, adjust, and maintain optimal pressure conditions.

1. **Start and Initialization**: The process begins with a "Start" command, followed by "Power ON," where the pressure control system is activated.
2. **Environment Scanning and Pressure Decision**: The system scans the surrounding environment to assess the current pressure levels and then determines the optimal pressure settings.
3. **Activation of Pressure Duct Control**: The system turns ON pressure duct control to adjust the pressure based on predefined scaling parameters.
4. **Monitoring and Adjusting**: With pressure duct control active, the system continuously monitors and if needed adjusts the pressure. When the pressure setting process is completed, it turns OFF the pressure duct control.

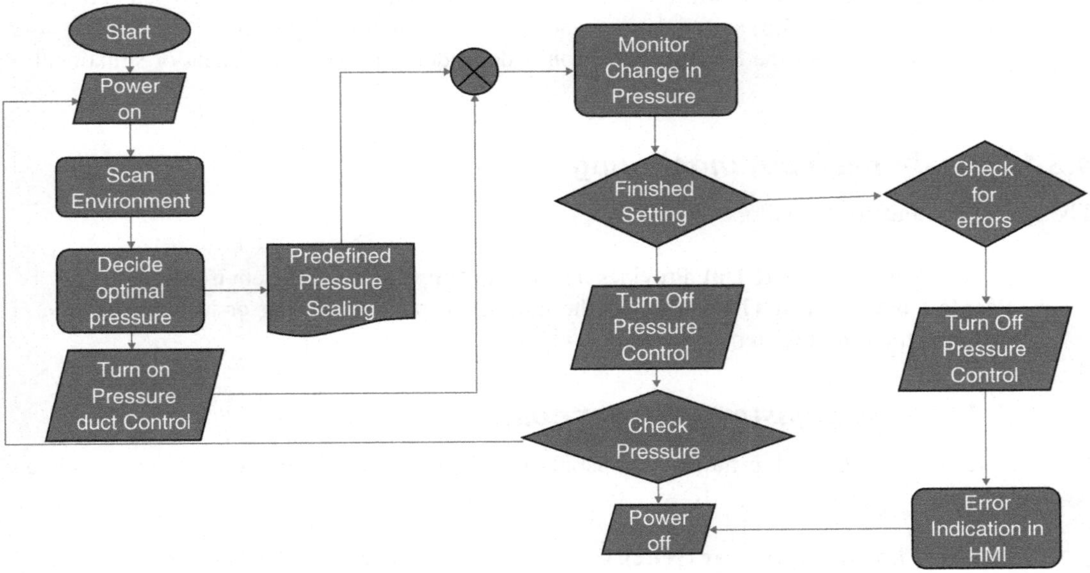

FIGURE 4.3 Algorithm for pressure control.

5. **Error Checking**: The system checks for any potential errors. If no errors are detected, it continues monitoring the pressure. If an error is found, the system shuts off the pressure duct control and provides an error indication on the HMI.
6. **Shutdown**: After successfully monitoring the pressure without any issues, the system proceeds to power OFF.

This process ensures efficient pressure regulation with an integrated error-detection mechanism to notify operators in case of issues. This setup is suitable for industrial or building environments where precise pressure control is critical.

4.3 RESULTS AND DISCUSSION

4.3.1 Working Principle of the Proposed PLC-Based System

The proposed PLC-Rockwell ControlLogix 5580-based system leverages temperature and people count sensors to dynamically adjust temperature control within an industrial environment. The system operates a combination of real-time inputs, internal registers, and predefined control logic to ensure optimal temperature regulation by controlling a heater and cooling fan. The stepwise working principle is as follows:

4.3.1.1 Initialization (base value setup)

- At the start of the program, the PLC initializes the base temperature thresholds and adjustment factor.
- The base low-set temperature (N7:0), base high-set temperature (N7:1), and temperature adjustment factor (N7:5) are predefined and loaded into internal registers.
- This system ensures the baseline operation is dependent on the environment or situational variations.

4.3.1.2 Real-time input monitoring

The system continuously monitors:

- **Temperature Sensor (I:1/0)**: Provides the current temperature of the environment.
- **People Count Sensor (I:1/1)**: Tracks the number of people entering or leaving the area, updating the N7:4 register.

4.3.1.3 Dynamic adjustment of set points

When there is a change in the number of people (N7:4), the adjusted temperature thresholds are recalculated:

- **Adjusted Low Temperature (N7:2)**: Calculated by adding the base low temperature (N7:0) and the current people count (N7:4) accounting for the adjustment factor (N7:5).
- **Adjusted High Temperature (N7:3)**: Similarly, the values are recalculated based on the base high temperature (N7:1) and people count (N7:4).

These dynamic set points ensure that the system adapts to changes in occupancy, thus maintaining the optimal temperature and pressure.

4.3.1.4 Heater control logic

The heater operation depends on the current temperature relative to the adjusted low temperature (N7:2):

- **Heater ON**: If the temperature falls below the adjusted low set point, the heater output (O:2/0) is energized.
- **Heater OFF**: If the temperature rises to or exceeds the adjusted high set point (N7:3), the heater output is de-energized.

4.3.1.5 Cooling fan control logic

The cooling fan is controlled based on the current temperature relative to the adjusted high temperature (N7:3):

- **Fan ON**: If the temperature exceeds the adjusted high set point, the cooling fan output (O:2/1) is energized.
- **Fan OFF**: If the temperature drops below the adjusted high set point, the cooling fan output is de-energized.

4.3.1.6 Real-time feedback and adaptation

1. The system continuously adapts to change in real-time inputs and controls output accordingly.
2. This ensures precise and dynamic control over temperature regulation, enhancing energy efficiency while maintaining operational stability in the industrial environment.

4.3.1.7 Key features

1. **Adaptive Control**: Automatically adjusts set points based on occupancy, improving responsiveness to environmental changes.
2. **Energy Efficiency**: Limits unnecessary operation of heating and cooling devices by maintaining optimized temperature thresholds.
3. **Reliability**: Operates with defined logic, ensuring consistent performance and minimal manual intervention.

The system's dynamic adjustment capability based on real-time sensor inputs makes it highly suitable for complex industrial applications requiring precise temperature management.

4.3.2 Deployment of PLC in Real-Time Industrial Applications

In this section, an advanced automation system is designed to monitor and control industrial temperature and pressure parameters with high precision and safety. Utilizing a PLC-Rockwell ControlLogix 5580 along with temperature and pressure sensors, this system automatically adjusts critical parameters in real time without human intervention. The PLC receives sensor data and processes it based on pre-programmed safety thresholds and executes adjustments as needed to maintain optimal levels. An

HMI provides real-time data visualization and remote control, allowing operators to monitor conditions, receive alerts, and make manual adjustments. This setup exemplifies how automation improves operational safety, accuracy, and efficiency, while minimizing risks associated with human error. The system is designed for reliability even in demanding industrial environments with built-in safety protocols that ensure consistent performance and safeguard both equipment and personnel.

4.3.2.1 Analytical results

In this section, the analytical results are derived from the system's performance with temperature and pressure sensors managed by PLC-Rockwell ControlLogix 5580. Through testing under simulated operational conditions, the system demonstrated robust responsiveness and accuracy in maintaining target parameters. Temperature adjustments were precise within a ±0.5°C tolerance, while pressure control maintained a ±2% accuracy across a range of operational levels. The PLC's processing capability proved effective in handling real-time data from multiple sensors and its response times met safety thresholds, confirming the system's ability to execute necessary adjustments promptly.

The HMI interface facilitated seamless operator interaction, allowing real-time monitoring and manual adjustments. Data collected over multiple test cycles indicated stable performance with minimal variance in output, showcasing the system's reliability. Overall, these results affirm that the automated control system can significantly enhance operational safety and efficiency, offering precise control and real-time adaptability in industrial applications.

4.3.2.2 Ladder logic

4.3.2.2.1 Temperature and pressure control

The ladder logic shown represents an automated control system for temperature and pressure management using a PLC. It continuously monitors sensor inputs and compares them against predefined thresholds using Less Than or Equal to (LEQ) and Greater Than or Equal to (GEQ) instructions to determine whether the values are within the acceptable range. Depending on the sensor readings, specific actions are triggered, such as activating outputs or internal control bits to adjust the system (e.g., cooling, heating, or venting). A MOV instruction is used to transfer sensor data to a register for processing, while a timer is employed to introduce delays in activating outputs, ensuring stable and precise control. This setup ensures real-time monitoring, efficient system responses, and human safety, making it ideal for industrial automation applications. Figure 4.4 shows the ladder logic code for the temperature and pressure function.

FIGURE 4.4 Temperature and pressure function.

4.3.2.2.2 Proximity sensor function

The ladder logic shown implements a proximity-based counting system using a Count Up (CTU) and Count Down (CTD) function. When a proximity sensor detects an object, the signal triggers the CTU to increment the counter, while another condition may activate the CTD to decrement the count. The preset value defines a threshold for actions based on the counter value. The Greater Than (GRT) instruction is used to compare the accumulated count with a set value (e.g., 100), triggering an output if the count exceeds this threshold. A Reset (RES) instruction is included to reset the counter to zero when specific conditions are met, ensuring the system operates within desired limits. This logic is typically used in industrial applications for inventory tracking, object detection, or machine cycle counting, ensuring efficient and automated monitoring. Figure 4.5 shows the ladder logic code for the proximity function.

4.3.2.2.3 Alarm excitation function

In the ladder logic for the alarm system using an **OR function**, the logic is designed to trigger an alarm output when any one of multiple fault conditions is detected. The OR logic is implemented by placing multiple input conditions (e.g., temperature threshold exceeded, pressure threshold exceeded, or system malfunction) in parallel branches. If **any** of these conditions become true, the circuit is completed, activating the alarm output.

This approach ensures that the alarm system is responsive to a variety of critical scenarios, improving safety and reliability in the operation. For example, if either a temperature sensor detects overheating or a pressure sensor identifies a dangerous spike, the alarm will sound immediately without waiting for additional conditions. Such a configuration is essential in industrial automation to prevent accidents and mitigate risks by alerting operators to take corrective action promptly. Figure 4.6 shows the ladder logic code for the alarm function.

4.3.2.2.4 Sprinkler function

In the ladder logic for the water sprinkler system using an AND function, the logic ensures that the sprinkler system is activated only when multiple specific conditions are met simultaneously. The AND logic is implemented by placing the input conditions (e.g., temperature exceeding a threshold and pressure within an acceptable range) in series. For the circuit to activate the sprinkler output, all conditions in the series must be TRUE.

This configuration is particularly useful for ensuring precise and controlled operation. For example, the water sprinkler system will only be activated when both a high temperature is detected (indicating a fire or overheating scenario) and the pressure in the water line is adequate for proper spraying. Such a setup prevents any false activation and conserves resources by ensuring that the system operates

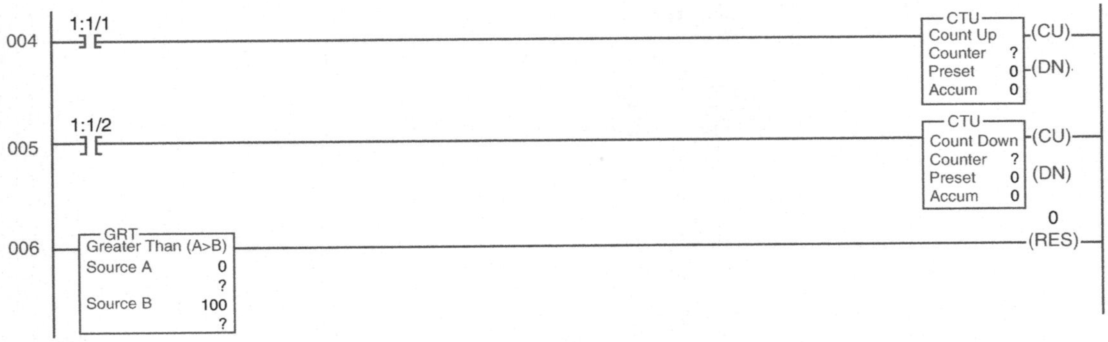

FIGURE 4.5 Proximity function.

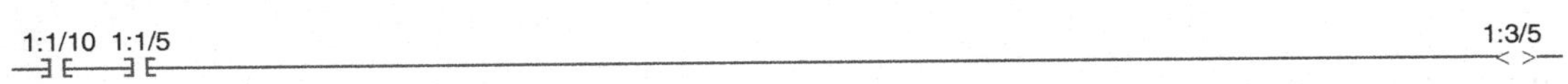

FIGURE 4.6 Alarm function.

only under the intended circumstances, enhancing the efficiency and safety of the sprinkler system in industrial or domestic applications. Figure 4.7 shows the ladder logic code for the sprinkler function.

4.3.3 Model Factory Setup

The image depicts a model factory setup designed to automate the monitoring and control of temperature and pressure parameters using PLC-Rockwell ControlLogix 5580 integrated with industrial-grade sensors and an HMI system. The PLC acts as the central controller, processing data received from sensors and automatically adjusting system parameters through actuators to maintain optimal conditions. The setup includes a real-time communication network that enables seamless data transmission between the sensors, PLC, HMI, and other systems, ensuring reliable operations. An HMI interface allows operators to monitor sensor readings, system performance, and alerts, while also providing options for manual control when required. Safety mechanisms are integrated into the system to trip operations during abnormal conditions, protecting both human operators and equipment. Additionally, a data logging system records operational parameters for diagnostics, analysis, and compliance reporting. This setup exemplifies Industry 4.0 principles, incorporating automation, safety, and real-time control to enhance efficiency and reliability in industrial operations. Figure 4.8 represents the model factory setup mentioned.

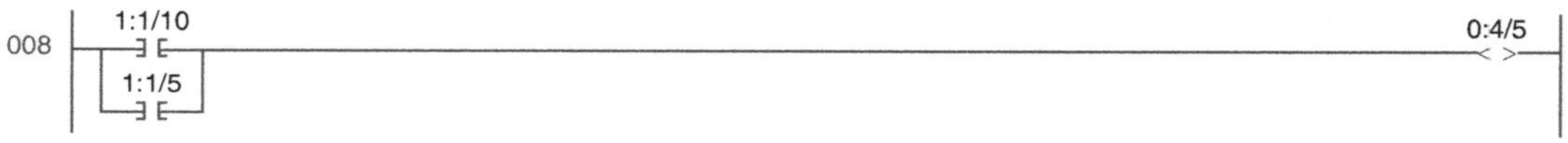

FIGURE 4.7 Sprinkler function.

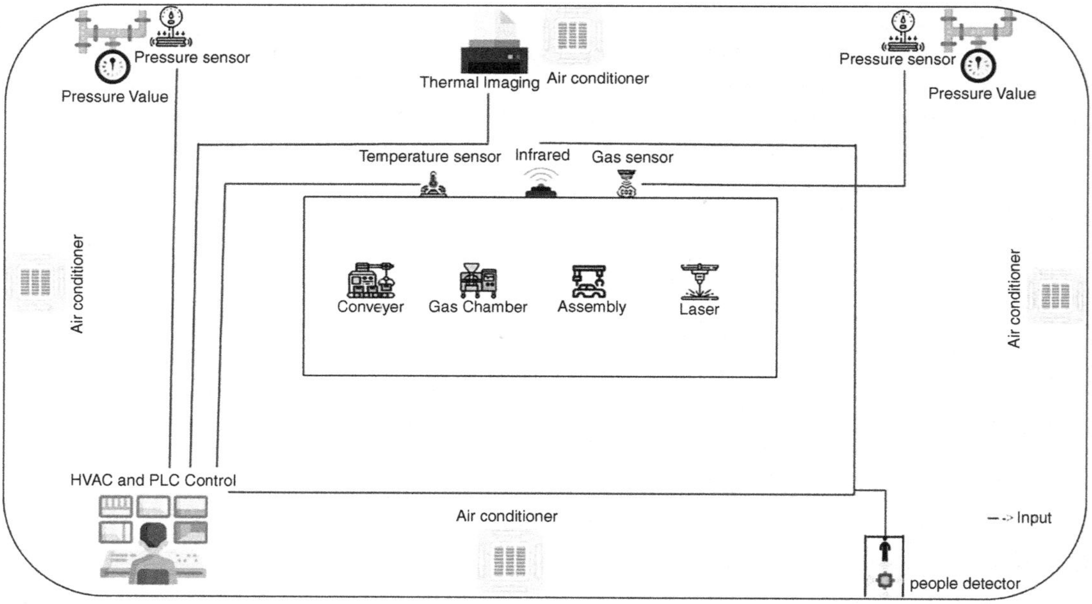

FIGURE 4.8 Model setup.

4.4 CONCLUSION

In conclusion, the development and integration of advanced PLC-Rockwell ControlLogix 5580 and HMI technology with pressure and temperature sensors present a significant leap forward in the field of industrial automation. This novel system autonomously adjusts critical parameters, eliminating the need for human intervention and enhancing operational efficiency and safety. By employing PLCs and HMIs, the system is able to constantly monitor and adjust environmental factors such as pressure and temperature in real time, allowing it to maintain optimal conditions with high precision. This not only increases the system's reliability but also minimizes the risks associated with manual control, which can be slow and prone to human error. The unique design overcomes many limitations found in traditional control systems, which often require manual oversight to operate effectively. By incorporating automation, this system achieves a level of adaptability that is essential in high-risk variable environments. In particular, the system's ability to autonomously stabilize parameters under changing pressure and temperature conditions demonstrates its suitability for a wide range of industrial applications, from manufacturing to energy production. For instance, in scenarios where rapid fluctuations in temperature or pressure could compromise equipment integrity or human safety, the systems' automatic adjustments mitigate these risks without the need for constant human supervision. This technology addresses key concerns within the context of modern industrial requirements, aligning with the broader principles of Industry 4.0. As industries embrace smart technologies, the demand for intelligent and interconnected systems grows. This system aligns perfectly with these trends by enhancing operational transparency and providing operators with real-time data through the HMI. The inclusion of intuitive HMI software allows operators to monitor the systems' performance from a safe distance, minimizing direct exposure to potentially hazardous conditions. In addition, the capability of the PLC to interpret sensor data and trigger appropriate responses makes this system a vital component in smart manufacturing environments, where downtime and safety risks must be minimized. Moreover, this system has significant implications for the security and resilience of industrial processes. As industries become more reliant on interconnected technologies, they are also increasingly vulnerable to cybersecurity threats. Integrating PLC and HMI technologies within the framework of secure and autonomous operations provides an added layer of safety by reducing human intervention, which can be a vector for accidental errors or intentional tampering. The system's robust design emphasizes automated, secure data handling and response mechanisms, ensuring stable performance even under potentially adverse conditions. This approach is also highly adaptable to different industrial settings. The modularity of the design allows it to be applied in various fields, such as power grids, smart cities, and automated production lines, where autonomous control systems are essential for maintaining safe and efficient operations. By combining adaptability with safety-focused automation, the system is well-suited for deployment in high-risk environments where human safety is a paramount concern. In summary, the combination of pressure and temperature sensors with PLCs and HMIs represents a comprehensive, forward-thinking solution for modern industrial challenges. This system's ability to autonomously monitor, adjust, and optimize critical parameters significantly enhances both operational efficiency and safety. Its alignment with Industry 4.0, adaptability across different industries, and focus on minimizing human exposure to risk make it a valuable asset in advancing safe and smart manufacturing practices. As industries continue to evolve, the principles embodied in this system will be essential in shaping a future where automation and safety go hand in hand.

4.5 FUTURE SCOPE

The scope of this project establishes a comprehensive foundation for implementing intelligent, autonomous control systems in industrial settings, aimed at optimizing both operational efficiency and safety. Through the integration of PLCs and HMIs, the developed system provides real-time monitoring and adjustment of critical parameters such as pressure and temperature, eliminating the need for constant human intervention. This capability not only ensures seamless and accurate control of these essential variables but also significantly enhances safety by reducing the likelihood of human error and promptly addressing hazardous conditions.

This project's applications extend across multiple industries—such as manufacturing, energy, chemical processing, and oil and gas—where automated control of environmental conditions is crucial. By autonomously managing these parameters, the system supports consistent operational stability, even under variable or challenging conditions, ultimately leading to improved efficiency, reduced downtime, and a safer working environment.

Furthermore, the project aligns with the objectives of Industry 4.0 by advancing digitalization and automation within industrial operations. The insights provided by this system enable proactive maintenance, allowing industries to anticipate and resolve potential issues before they impact operations. This solution contributes to the ongoing evolution of industrial safety and automation, establishing a new benchmark for autonomous safety-driven control systems and underscoring the value of integrated PLC-HMI frameworks in creating resilient, high-performance industrial environments.

REFERENCES

1. Ouberri, Y., Yatimi, H. and Aroudam, E., 2020. Design of a robust sliding mode controller for MPPT based on automation PLC for PV applications. *International Transactions on Electrical Energy Systems*, 30(4), p. e12296.
2. Cui, X., Kou, Z. and Qiao, Y., 2019. Water-level control system for solar water heating engineering based on PLC. In: *2019 IEEE 3rd Information Technology, Networking, Electronic and Automation Control Conference (ITNEC)*, Chengdu, China, pp. 1625–1628. IEEE.
3. Toroman, A. and Mujčić, E., 2020. Design and experimental analysis of the smart traffic control system based on PLC. In: *Advanced Technologies, Systems, and Applications IV-Proceedings of the International Symposium on Innovative and Interdisciplinary Applications of Advanced Technologies (IAT 2019)*, Sarajevo, Bosnia and Herzegovina, pp. 470–480. Springer International Publishing.
4. Mahoor, M., Hosseini, Z.S., Khodaei, A., Paaso, A. and Kushner, D., 2020. State-of-the-art in smart streetlight systems: a review. *IET Smart Cities*, 2(1), pp. 24–33.
5. Fathahillah, F., Siswanto, M., Fauziyah, M., Parlindungan, R., Putri, R.I. and Roh, Y.G., 2020. Implementation of programmable logic controller in multi machine operations with product sorting and packaging based on colour detection. *IOP Conference Series: Materials Science and Engineering*, 732(1), 012069.
6. Hassan, S., Yusof, M.S., Embong, Z., Kamarudin, Q.E., Haq, R.H.A., Ibrahim, M., Khairilhijra, K.R., Marwah, O.M.F., Sa'ude, N., Ismon, M. and Madlan, M.A., 2019. A study of frequency and pulses for stepper motor controller system by using programmable logic controller. *Journal of Physics: Conference Series*, 1150(1), 012040.
7. Chan, C.F., Chow, K.P., Mak, C. and Chan, R., 2019. Detecting anomalies in programmable logic controllers using unsupervised machine learning. In: *Advances in Digital Forensics XV: 15th IFIP WG 11.9 International Conference, Revised Selected Papers* 15, January 28–29, 2019, Orlando, FL, USA, pp. 119–130. Springer.

8. Rezig, S., Ghorbel, C., Achour, Z. and Rezg, N., 2019. PLC-based implementation of supervisory control for flexible manufacturing systems using theory of regions. *International Journal of Automation and Control*, 13(5), pp. 619–640.
9. Sheela, A., Shibi, S.P., Vinothkumar, S., Prasanna, V. and Jagadeeshwaran, A., 2023. PLC based levelling and positioning control system. In: *2023 9th International Conference on Advanced Computing and Communication Systems (ICACCS)*, Coimbatore, India, pp. 1262–1265. IEEE.
10. Al-Andzar, M.F. and Puriyanto, R.D., 2019. PID control for temperature and motor speed based on PLC. *Signal and Image Processing Letters*, 1(1), pp. 7–13.
11. Kumar, D., Basit, A., Saleem, A. and Abbas, E.G., 2019. PLC based monitoring and protection of 3-phase induction motors against various abnormal conditions. In: *2019 2nd International Conference on Computing, Mathematics and Engineering Technologies (iCoMET)*, Sukkur, Pakistan, pp. 1–6. IEEE.
12. Zhao, H. and Zheng, J., 2020. Design and research on temperature control system of electric furnace based on PLC touch screen integrated machine. *IOP Conference Series: Materials Science and Engineering*, 914(1), 012034.

Healthcare Sector *Online Oxygen Cylinder Management System*

5

V. Anantha krishnan, Tanya, Dishank Singh, Andrila Priyam, R. Lalitha Kala, Harish Raaghav, and O. V. Gnana Swathika

5.1 INTRODUCTION

The human body needs oxygen to thrive and perform a variety of important functions. To deliver oxygen to the body, the ventilator works in conjunction with the circulatory system. When oxygen is inhaled, it moves to the lungs and then to the blood arteries, where it mixes with the blood. Next, oxygen-rich blood is pumped into the cells and tissues of the body. However, some people who are subjected to health constraints develop hypoxemia as a result of lung inflammation and infection (lower blood oxygen levels) [1].

These individuals require medical oxygen to restore the body's oxygen saturation. Medical oxygen is a very pure form of oxygen that is used to enhance oxygen levels in people suffering from critical illness such as chronic obstructive pulmonary disease (COPD), cardiac or respiratory arrest, shock, cyanosis, and severe bleeding. Medical oxygen is delivered by a vacuum-insulated evaporator (which keeps liquid oxygen cool), an oxygen cylinder, or an oxygen concentrator (separates oxygen from the atmospheric air). SARS-CoV2, a new coronavirus, binds to the ACE2 receptors found in the lungs. The virus affects the epithelial cells that line the respiratory system and protects it from infections and diseases [2].

When people become infected, their bodies go through a series of changes: the oxygen and carbon dioxide exchange in the alveoli (tiny sacs inside the lungs) is disturbed. This contamination activates the immune system, which responds by releasing immune cells such as cytokines and interleukins to combat the infection. When this immune response persists for a long time [3], it causes inflammation and further obstructs oxygen flow inside the lungs. Humans with severe health constraints may develop edema and fluid build-up in the lungs as a result of repeated exposure, eventually leading to pneumonia. Acute respiratory distress syndrome (ARDS), which is characterized by extreme dyspnea and a fast respiratory rhythm, can also be caused by severe damage to the alveoli [4,5].

DOI: 10.1201/9781003649700-5

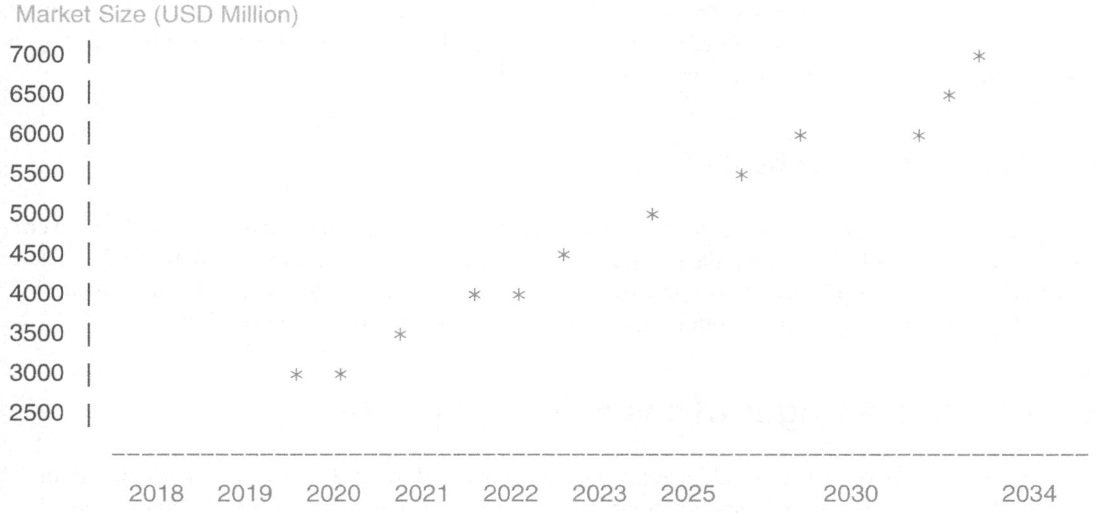

FIGURE 5.1 Medical oxygen cylinders market.

All of these situations require oxygen treatment and can make it difficult for the patient to breathe normally. Figure 5.1 shows the US Medical Oxygen Cylinders Market projections. To cope with the current situation, you need an effective online Oxygen Management System.

5.2 MOTIVATION

As health crises are spreading globally and the main challenge is soaring demand for oxygen cylinders, it is hence necessary to provide a novel Oxygen Cylinder Management System. The proposed system is a bid to enhance the capability to fulfill the surprising rise in the demand for medical oxygen and health center beds globally. Apart from oxygen generation, the health crisis instances surged in distinct elements of the country, leaving infrastructures in hospitals under vast stress. Our project focuses on generation of information of lenders and borrowers who are in need of oxygen and helping them to reach out to receive the required support.

5.2.1 Objectives

The goal of this venture is to create an oxygen cylinder management device with the intention to offer a platform for all the users to know approximately the supply of oxygen cylinders of their location. They can borrow or donate their oxygen cylinder in times of need.

5.2.2 Scope of Study

Oxygen Cylinder Management System is a system that is designed to keep, manage, retrieve, and examine statistics related to the executive and inventory control inside a specific city. Our proposed undertaking aims at retaining all the data relating to oxygen cylinder donors and the availability of oxygen cylinders in every town and assist them to manage this information more effectively. Our proposed undertaking aims to retain all data relating to oxygen cylinder donors and the availability of oxygen cylinders in each town, and to help manage this information more effectively.

5.2.3 Existing System

The second wave of coronavirus created unexpected challenges around the world as hospitals in several parts of the world, including Delhi, the capital of India, suffered from severe oxygen deficiencies. There was no longer a suitable channel to connect the oxygen supply. Existing systems often communicate via phone, SMS requests, or posting requests to social media sites such as Facebook, Twitter.

5.2.4 Disadvantages of the Existing System

- The majority of district-level hospitals lack a centralized medical gas system with piping. In addition, many county hospitals in large cities rely on distributors, despite the high demand for medical-grade oxygen. Therefore, a review of the oxygen supply system is needed to ensure an uninterrupted oxygen supply in all district-level hospitals.
- The wrong approach prevented them from meeting the needs of the victims, resulting in many deaths.
- There is a communication delay.
- The lack of scientific oxygen has ruined the health of many patients, resulting in their death.

5.3 PROPOSED OXYGEN CYLINDER MANAGEMENT SYSTEM

In the current scenario, it is necessary to find an oxygen distributor in a specific location. Many have died because they cannot find an oxygen supplier in their area. We have created a solution. What if you had a website where you could enter your login information and see where you can buy oxygen in your area and its availability?

5.3.1 Proposed System and Its Advantages

The block diagram of the proposed model is shown in Figure 5.2. During these dangerous times, an uninterrupted supply of oxygen to hospitals is critical for the safe operation of medical facilities. The platform will show the location of assigned transportation with the undertaking of oxygen delivery, and will assist in providing oxygen to hospitals in the quickest possible period based on demand and necessity.

The suggested equipment might be used to ensure that public and private hospitals have access to oxygen 24 hours a day, seven days a week. The needy will be able to communicate directly with officers/people in control of the state's oxygen supply chain via the internet link. By way of the employer's representatives, all of the facts may be kept up to date on the portal on a regular basis by checking its availability online.

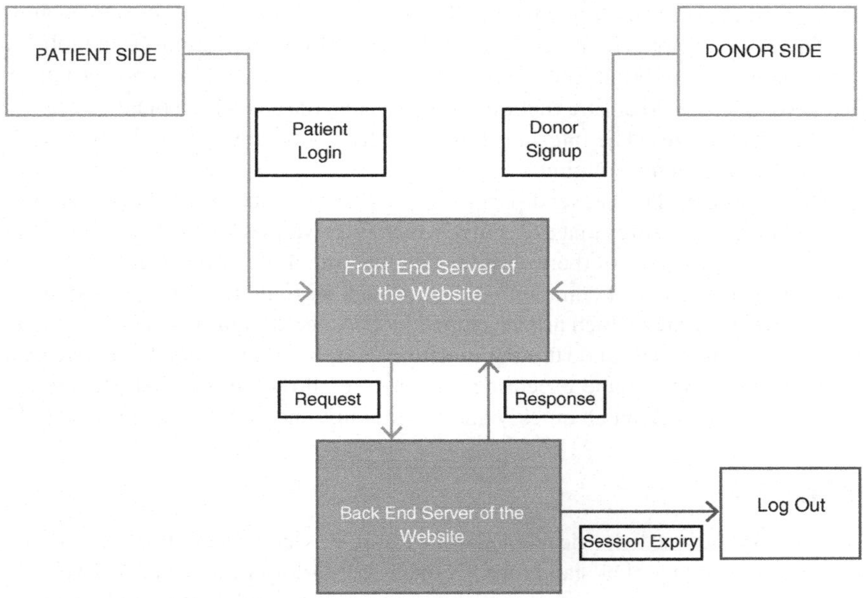

FIGURE 5.2 Block diagram of our proposed model.

The advantages are as follows:

- The proposed method of providing oxygen is uncomplicated and timely.
- Prevention of death in the event of a need or availability of an oxygen supply.
- A particular nursing home that wants to increase oxygen availability should better understand the unique oxygen delivery options and solutions available.
- An overview of oxygen demand and oxygen delivery options in a single site or area uniformly required by a site or area.
- In the event of unavailability or in the event of a serious emergency, persons in need may immediately contact the relevant service provider.

5.3.2 Software Used

- **Front end**: HTML, CSS
- **Backend**: Python

Now we can have a detailed explanation about each of the above-mentioned software tools.

Front End and Backend: The famous terminologies used in web development and web-based services are front end and backend. These terminologies are considered as important for network improvement, but they both are quite different from each other. To make the website work better, each element wants to interact and work well with the others as a unit.

Front-End Development: User interface refers to the elements of a web page that the user interacts with immediately. It is also called the "buyer" of public services. It includes everything the customer sees and interacts with, like text, colors and patterns, images, charts and tables, buttons, colors, navigation menus. The languages used for Front-End development are HTML, CSS, and JavaScript.

These respective Front-end developers provide the basic structure, layout, behavior, and content of everything displayed on a computer screen when websites, online applications, or a mobile application is deployed. The main goals of Front End are responsiveness and the improved performance it delivers. The respective developer should ensure that the website is responsive, which means it should work well on devices of all different sizes. The interface renders correctly on devices of all sizes, with consistent functionality across any screen resolution.

Backend Development: The backend part of any website is considered the server side. It records and organizes statistics, and ensures that every task which is forwarded by the client side of the website is running smoothly. This is the part of the website which we cannot see or interact with. It refers to a part or a piece of software that does not come into direct contact with the customer. Consumers can access elements and respective features which are developed by backend designing team. Writing and creating new APIs, building new libraries, and running machine components without respective user interfaces or even a scientific programming and executing systems are all on the backend. Python is a programming language that lets us work much quickly and successfully integrate the systems into the backend.

5.3.2.1 HTML

Tim Berners Lee was the one who developed HTML. It was composed for the first time in 1993. HTML is a computer language that stands for "HyperText Markup Language." HTML is made up of several elements [2]. The first is Hypertext, which refers to a "clickable" Markup Language method of symbolizing or implying information. To summarize, HTML is a computer language that incorporates hypertextual material that may be clicked on. In other words, HTML is the engine of web development. The site layout can be viewed on any web browser. Accordingly, the complete structure of a web page in HTML code is suggested by this site proposal. The code is used to build the HTML.

These codes provide the web browser with the most efficient instructions on how to display or compose content. We cannot see the backend code of an internet website page. The backend encoding can be found on every page of a website. You can easily view this code by going to > Preferences > View Licensing Site. In other words, this page contains all written or proven documented instruction code. As a result, textual content is written on various internet pages. The instructions are written in the backend, and the text is shown in front of us. HTML code may be read by any web browser. It reads the HTML code and generates a result.

Furthermore, the HTML coding is a fundamental building component of any internet site or internet-related applications. A website is made up of many collections of web pages and a website is made up of several web pages. Each page is generated with its own code. So there are three main components of a website. First of all, we must not ignore the HTML. Second, there's Python, then there's CSS.

The importance of each element in web development cannot be overstated. Something is incomplete if it has no alternative. However, when we consider the importance and power of HTML, we can see that HTML, as the name suggests, is capable of handling text documents. CSS is the design of the text document, while Python is its function. In addition, HTML together with its partners ensures that a website is both useful and attractive. Sometimes we come across text elements like "click here." When you click on it, you will be redirected to a new page. This is what super engineering achieves. HTML has an important role in web development. HTML, for example, strongly

- Defines the size of the text.
- Defines the form/font style of the textual content.
- Aligns the textual content in exceptional portions/directions.
- Arranges the textual content in tabular form.
- Refers web page visitors to the following web page.
- Supports snapshots so you can show them on the net web page.

- Supports films too.
- With the help of Java, it performs calculations and performs conditional duties.
- Can embed distinct scripts in itself to perform complicated obligations.

5.3.2.2 CSS

The essence of a website is its web design. It is essential for the success and performance of a website. This is why business website design is of particular interest to web developers. A well-designed website will attract more visitors and rank higher in search engines.

CSS stands for Cascading Style Sheets, which has greatly simplified the creation of web pages. CSS allows you to easily create links to different files on a web page. CSS allows you to control different elements on different web pages of your website. The shape and manner of displaying content on a web page is determined by proprietary CSS. It has nothing to do with how a website is designed. On the other hand, CSS will affect the look and feel of the layout when the final engineering is complete. With the use of a CSS sheet, you may manage the font, placement, color, and style statistics of a whole website. It enables us to improve the overall look and haptics of a particular website. When new changes are made to a single stylesheet, CSS allows those new changes to cascade or ripple across all of the sheets which are attached to it, much like a waterfall's cascading or in other terms ripple effect. While HTML is concerned with the content and the basic structure of a website, CSS focuses on the presentational components such as attractive layout, typeface, and color. Most of these factors had to be protected in the HTML itself before CSS was introduced, which made the documentation part more difficult. This separation of various different functions between HTML and CSS allows the respective website designers to have more control and concreteness toward the web pages.

CSS also allows the same web page to be displayed differently on different media, such as printers, monitors, or possibly touch-enabled devices. It also determines the sound of a document when spoken aloud. Thus, the writer can change the amount, sound of history, and many other aspects of a report before it is used in the spoken medium when building a web page using CSS.

CSS is user-friendly and interoperable with the majority of internet browsers. It allows you to easily place and rearrange elements over a web page online. The highly important feature of CSS, which makes it so important in web design, is that it emphasizes on layout above content. CSS allows an internet designer to easily modify and manage all aspects of a website's formatting, which greatly facilitates the work of an online designer.

The following are the advantages and uniqueness of CSS over other languages:

Quick and Fast Loading with Low Bandwidth Consumption: CSS is chosen by the majority of the web designers because it is lighter than the table layout, which requires a lot of bandwidth to process and deliver. The style sheet is just loaded once and it gets saved in the cache memory. The future pages will load in a fast manner as a result of this.

Multiple Browser Compatibility: CSS is well suited to all web browsers. As a result, a website built with CSS will look the same across all browsers.

Ability to Position Elements Anywhere on the Page: CSS allows you to position elements anywhere on the page. As a result, if the location of a certain detail does not flow well with the way you wanted it to, you may ask the developer to swap it out.

Print-Friendly Web Pages: Another important characteristic of CSS is that it allows for the creation of print-friendly web pages. This means that the web pages are easily accessible. Colorations, photographs, and other difficult to print items may all be easily deleted and published.

Customization: CSS sheets enable clients to personalize a web page. The CSS sheets are kept outside and allows the respective user to make whatever modifications they see necessary. Users of a variety of cutting-edge browsers have the option of creating their own fashion sheets. As a result, a person can easily exchange font over homes and other items.

Search Engine Friendly and Optimized: CSS style sheets make it much easier for a website to appear in crucial search engine results. Because it allows items to be placed anywhere, the user may project the most important material first, ensuring that web crawlers can much easily catch it.

Consistency: CSS ensures that web pages are all the same. It means that all expressions and texts will inherit their characteristics from an external or outsourced style sheet. As a result, a respective developer does not have to be concerned about the characteristics of the factors changing because they can be easily changed at any of the ongoing stages of web development using CSS.

CSS does contribute toward the mobility of the content material. You can create separate sheets for different media types using CSS. Therefore, materials can be transferred flexibly. It also allows you to change the quality of the variables on the site to suit the needs of the situation.

5.3.2.3 Python

Guido van Rossum first invented Python in 1991. His ideology emphasizes code readability, which is demonstrated through basic syntax, namespaces, and indentation tension. Python's ease of use and readability makes it a popular choice as an introductory language for many people. Python is a simple yet powerful programming language that can be easily run on most popular frameworks (Windows/Linux/Mac). The language can follow a programmer's coding style (functional, task-oriented, or article-oriented), allowing developers to choose the approach that best suits their current work.

Internet development deals with the front end (everything that interacts with the user) and the backend (which is hidden from the normal user) that contains all the business intelligence and interacts with the database. Python is used as a backend language in web development, and it is often combined with another front-end language (usually JavaScript) to create a complete web page.

The reason for Python's popularity in web development is simple: it is a highly adaptive, flexible, and efficient programming language with dynamic typing capabilities. Python allows programmers to create clinical applications, device programs with images, games, command line tools, web applications, and more. In fact, Python has long been considered the "most popular coding language" in the industry. Another big motivation for adopting Python is that it's completely free of cost to use and is distributed by the end user manifold. Python is an open-source programming software or product that works with a greater number of libraries. Since all the necessary coding information is available over the internet, there is no limit to copying, integrating, and distributing Python for your product.

There are a few things that make it easy to use Python for networking:

Can Be Understood Simply: The simplicity of the syntax makes it possible to handle complex systems and ensures that communication between developers working on the same project becomes much greener. In fact, an easy language to learn means that novice developers or those with no Python experience can learn the language and join the development team sooner.

Good Readability: The fact that Python is similar to our native language and the emphasis on readability means that its creators can produce code that is easy to understand.

Complex Backend Tasks: Python is a challenging coding language that allows you to perform complex backend tasks like artificial intelligence (AI) and data science, as well as other common tasks in language programming.

High Reputation among Internet Improvement: This attribute is generally considered an indication of the suitability/quality of a language, but is also useful for debugging our application. If you're having problems with a malicious app or you're not sure how to implement a new feature, chances are someone on the internet is having the same problem and it's only a few clicks away.

Wide Range of Libraries: You can use a wide selection of Python libraries to speed up your utility development. These libraries are pre-written code programs that anyone can use, eliminating the need to recreate the wheel. Numpy, scitLearn for information evaluation and math methods, and x for composable SQL queries are just a few of the libraries available.

Django, Flask, Pyramid, Web2Py, and Turbogears are just a few of the most popular network enhancement frameworks for Python. These frameworks are similar to toolkits in that they try to speed up building a web widget. They include standardized code packages and modules that can help you with various software areas (URL routing, database access, HTTP requests and responses).

5.3.3 Functionalities, Modules, and Working

The modules used are:

- Donor
- Patient
- Admin

Donor:

1. Donors can create an account by submitting basic information.
2. Donors can contribute oxygen tanks after logging in, and oxygen tanks will only be put into storage after getting permission from the admin.
3. Donors can view their donation history as well as their current status (pending, approved, declined).
4. Donors can also choose from oxygen cylinder supplies.
5. Donors can view the status of their oxygen tank claim.
6. On their dashboard, donors can see the number of oxygen tank requests that have been made, approved, pending, and declined by the admin.

   Note: Donors have the option of donating oxygen cylinders or requesting oxygen tanks.

Patient:

1. Create account (no approval required by admin, can login after signup).
2. After Login, can see number of oxygen cylinder request made, approved, pending, rejected by admin on their dashboard.
3. Patient can request for oxygen cylinder of a specific city from stock.
4. Patient can see their oxygen cylinder request history with status (Pending, Approved, Rejected).

Admin:

1. Admin should login to the profile via the username and credentials.
2. Once they log in, they will be able to access the dashboard where they can see the overall availabilities of oxygen state wise.
3. Another important work of the admin is to approve the credible request which are raised by patients and donors.
4. They keep a track on oxygen demand and availability of stocks.
5. They maintain the overall website.

5.4 WEBSITE SNIPPETS

Figure 5.3 shows the homepage of our proposed Oxygen Cylinder Management System. Figure 5.4 shows how the patient can request for oxygen cylinders. Figure 5.5 shows the individual blood request and approval page of the patient login. Figure 5.6 shows the donor login page and where they can donate oxygen cylinder. Figures 5.7 and 5.8 show a bird's-eye view of the admin page of the proposed system.

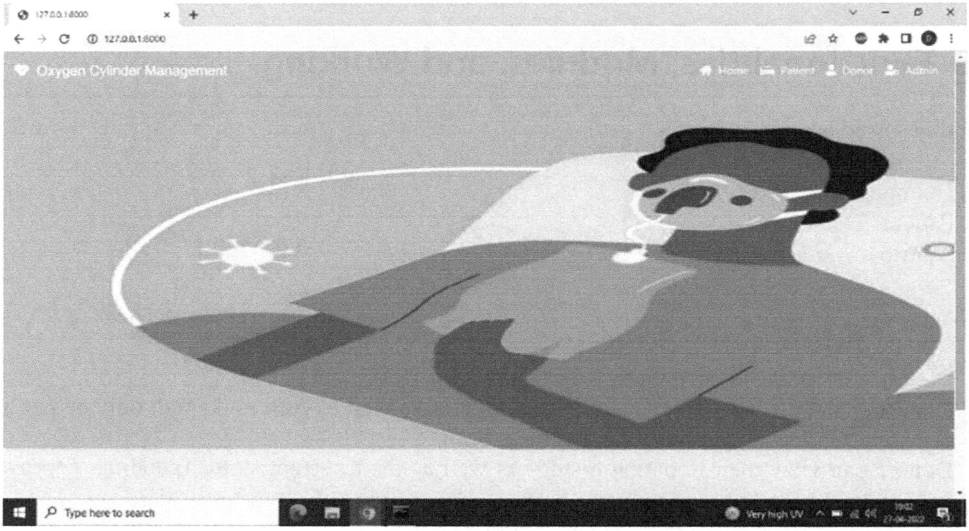

FIGURE 5.3 Home page of our Oxygen Cylinder Management System.

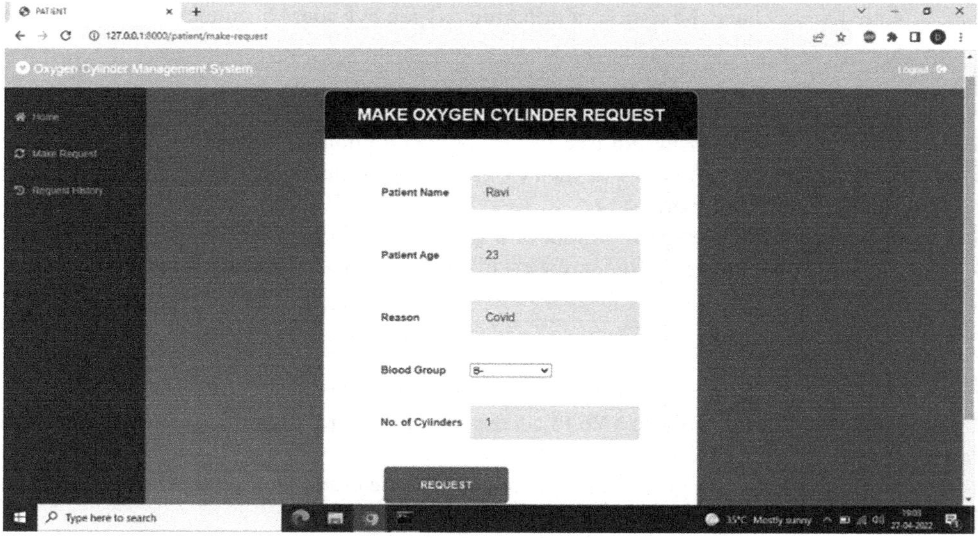

FIGURE 5.4 Patient request page of our Oxygen Management System.

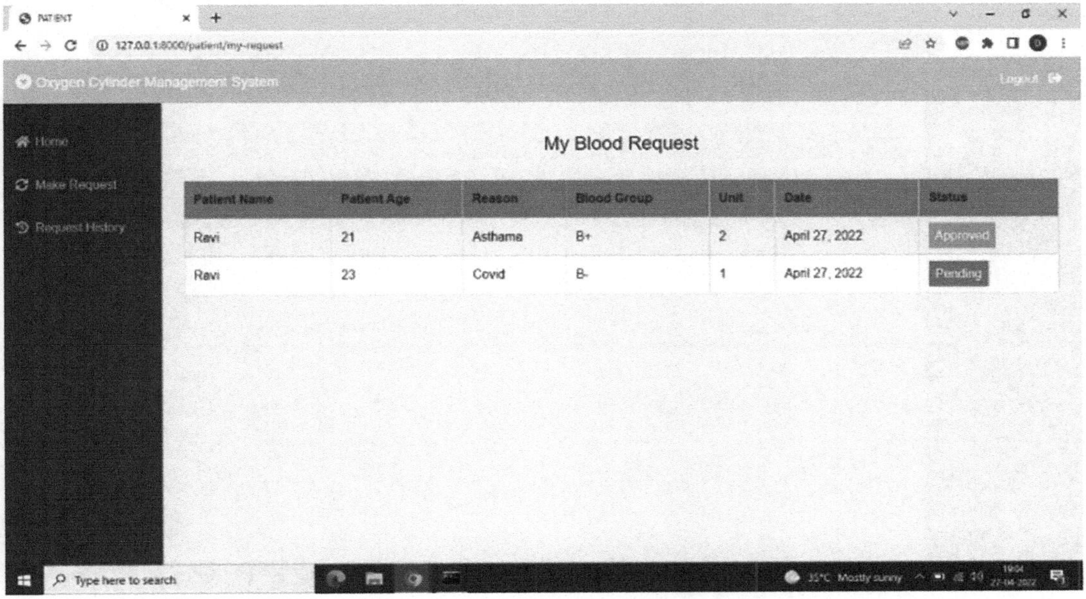

FIGURE 5.5 Individual blood request and approval page.

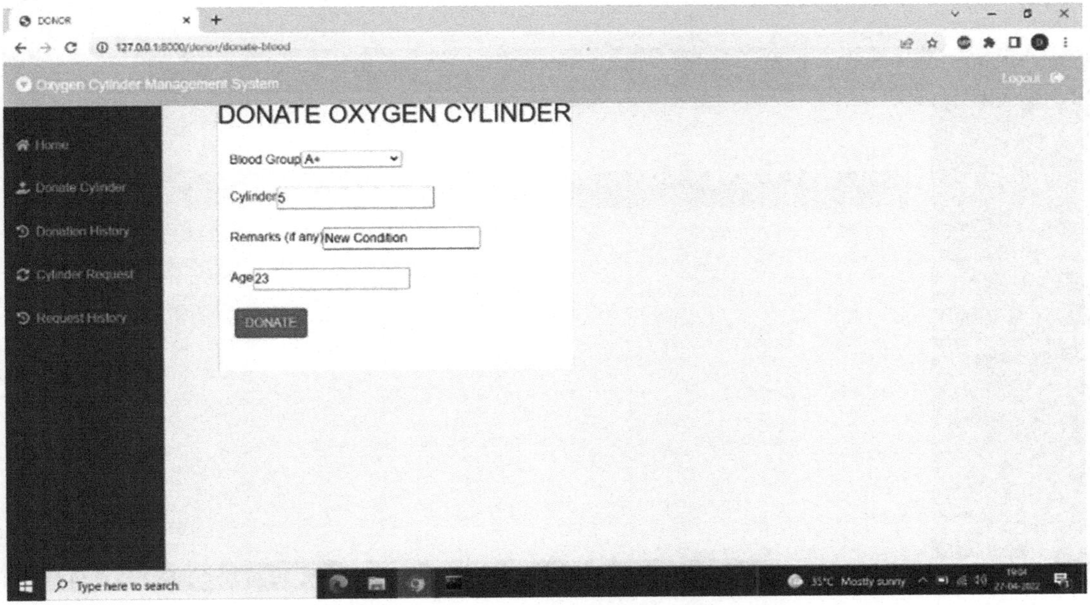

FIGURE 5.6 Donor login with which they can donate the oxygen cylinder for the patient login.

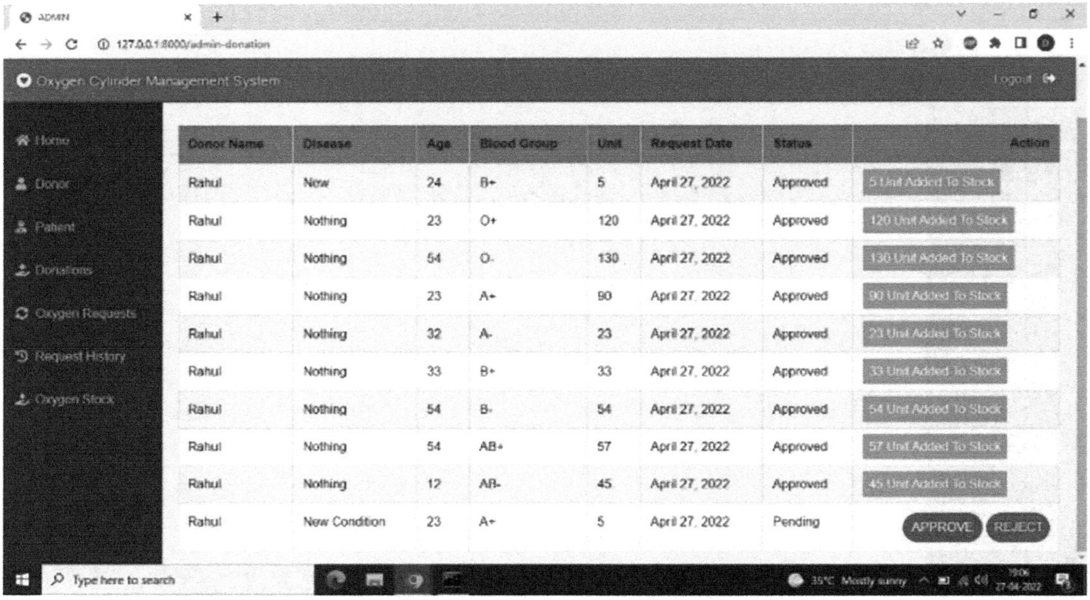

FIGURE 5.7 Admin login and inventory management of oxygen.

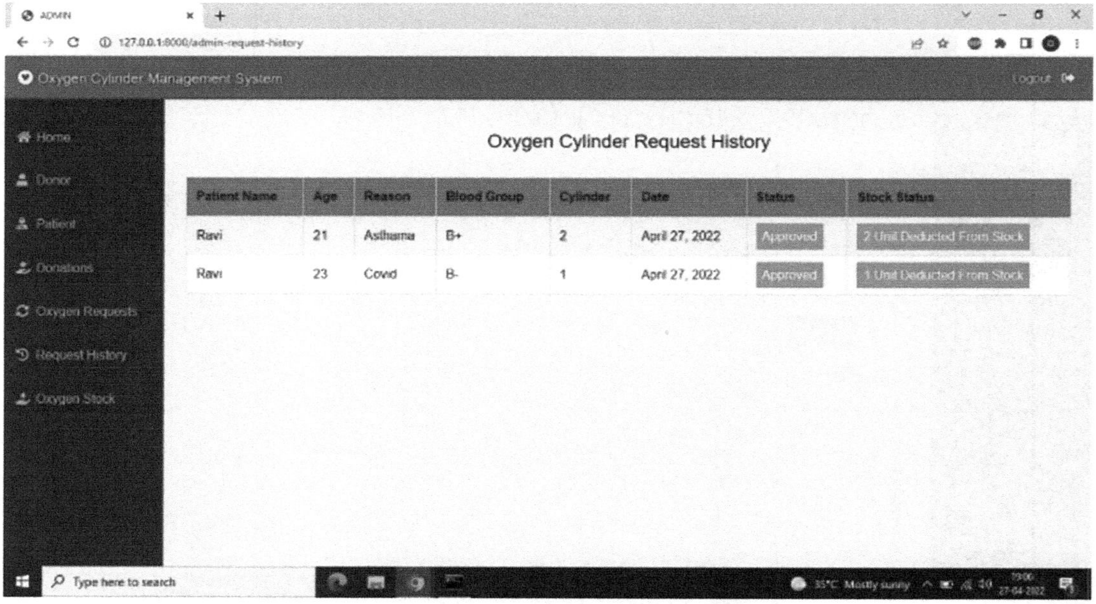

FIGURE 5.8 Admin oxygen cylinder request history.

5.5 CONCLUSION

Thus, we have created an oxygen cylinder management website with the intention to offer a platform for all the users to know approximately the supply of oxygen cylinders of their location. Based on this, they can borrow or donate their oxygen cylinder in times of want and our Oxygen Cylinder Management System is a system this is designed to keep, manage, retrieve, and examine statistics related with the executive and inventory control inside a specific city. Our proposed undertaking retains all the data relating oxygen cylinder donors, oxygen cylinders to be had in every town, and assist them manipulate in a better way. It provides transparency in this area, making the technique of obtaining oxygen cylinder trouble free and without corruption and making the system of oxygen cylinder control effective. This proposed technology will be released as a digital platform to accelerate both individual and hospital supply chains to combat the tragedy of oxygen deficiency in the face of the raging health crisis.

REFERENCES

1. Vignesh, C., Jagadesh, R., Ashifa, K., Subesh, N. R. and Kiruba Devi, T., 2023. Developing efficient methods for tracking the location, status, and maintenance needs of oxygen cylinders equipped with IoT devices. In: *International Conference on Internet of Everything and Quantum Information Processing*, Vijayawada, Andhra Pradesh, India, pp. 170–175. Cham: Springer Nature Switzerland.
2. Blakeman, T.C. and Branson, R.D., 2013. Oxygen supplies in disaster management. *Respiratory Care*, 58(1), pp. 173–183.
3. Bikkina, S., Manda, V.K., Rao, U.A. and Prasadarao, S.S., 2020. Are oxygen gas cylinders safe for home medical, usage? *Journal of Critical Reviews*, 7(15), pp. 3583–3589.
4. Graham, H.R., Bagayana, S.M., Bakare, A.A., Olayo, B.O., Peterson, S.S., Duke, T. and Falade, A.G., 2020. Improving hospital oxygen systems for COVID-19 in low-resource settings: Lessons from the field. *Global Health: Science and Practice*, 8(4), pp. 858–862.
5. Howie, S.R., Ebruke, B.E., Gil, M., Bradley, B., Nyassi, E., Edmonds, T., Boladuadua, S., Rasili, S., Rafai, E., Mackenzie, G. and Cheng, Y.L., 2020. The development and implementation of an oxygen treatment solution for health facilities in low and middle-income countries. *Journal of Global Health*, 10(2), 020425.

6 FPGA-Based Solar-Powered EV Charging Station: *Smart Infrastructure for a Sustainable Future*

S. Nakul, S. K. Preetham, R. Pradeesh,
K. Kartheesan, K. T. M. Udayanga Hemapala,
and O. V. Gnana Swathika

6.1 INTRODUCTION

Electric vehicle (EV) charging infrastructure has evolved significantly from slow residential alternating current (AC) chargers to high-power direct current (DC) fast chargers. Traditional EV charging systems typically use microcontroller or digital signal processor (DSP)–based control to manage power electronics and user interfaces. The rapid growth in EV adoption has brought new challenges including grid overloading during peak demand, the need for real-time load balancing, and the importance of maintaining battery health through precise charging algorithms. Integrating solar photovoltaic (PV) systems into EV charging infrastructure not only supports sustainability goals but also reduces reliance on the conventional power grid. A typical solar-powered EV charging station incorporates PV panels, DC-to-DC converters equipped with Maximum Power Point Tracking (MPPT), Energy Storage Systems, and power electronics interfaces. When advanced MPPT algorithms are implemented on reconfigurable hardware platforms like field-programmable gate arrays (FPGAs), it enables dynamic adjustment of converter duty cycles to maximize solar energy capture even under varying irradiance and temperature conditions. FPGA-controlled MPPT improves convergence speed and reduces energy

DOI: 10.1201/9781003649700-6

loss compared to microcontroller-based systems. Hybrid storage systems such as batteries combined with supercapacitors buffer solar intermittency and enhance fast-charging capability. FPGAs are utilized for effective energy and load management, real-time monitoring and control within the charging infrastructure. They offer advantages such as fast response times, parallel processing, and flexible system control which are crucial for dynamic energy management [1]. Specifically, FPGAs are programmed to dynamically allocate power based on the state of charge (SoC) of the EV battery and user needs. In hybrid station designs, multiple energy sources or storage are combined. For instance, a hybrid energy storage system might include batteries plus supercapacitors to buffer PV intermittency and fast load changes. Power management algorithms can run on FPGAs to decide when to draw from PV, battery, or storage. Wireless power transfer (WPT) is also being explored where an FPGA controls resonant inverters for contactless charging. Early prototypes demonstrate that an FPGA can generate optimized pulse-width modulation (PWM) waveforms for a static wireless EV charger which ensure safe power flow over moderate coil misalignments. FPGAs offer parallel processing, low latency, and hardware-level reconfigurability, making them ideal for real-time control in EV charging stations. Key applications include MPPT, Battery Management Systems (BMSs), and power electronics control for bidirectional DC-DC converters and inverters. FPGA-based controllers enable dynamic switching between charging modes (constant current/constant voltage), fault detection, and rapid response to fluctuating grid and load conditions. Research demonstrates improved efficiency, power quality, and system reliability in FPGA-based systems compared to traditional controllers. For instance, FPGA-implemented reinforcement learning (RL) algorithms dynamically allocate charging currents across multiple EVs balancing grid constraints with user priorities.

DC architectures prevail because they circumvent stage conversions inherent in AC fast chargers and enable direct DC-to-DC coupling between PV strings, battery banks, and vehicle packs. High-frequency (>100 kHz) isolated dual-active bridge converters controlled via phase-shift PWM in FPGA fabric furnish galvanic isolation and bidirectional flow for emerging vehicle to grid (V2G) services. Designs featuring synchronous buck or multiphase boost legs achieve >20 kW per module with current ripple under 2%. The use of Three-Port Converters (TPCs) especially with PWM and Phase-Shift Controllers is proposed to reduce switching losses and improve reliability, achieving high efficiencies [2]. Thermal considerations are mitigated through SiC MOSFET adoption and adaptive dead-time control implemented in PWM state machines. At the heart of a solar EV charger are power electronic converters. DC-DC stages interface PV arrays to batteries or chargers and DC-AC inverters (for AC output or wireless coils) drive the vehicle interface. FPGA control can be embedded directly into these converters. For instance, FPGAs allow concurrent execution of advanced control algorithms with power electronic hardware like bidirectional DC-DC converters. In practice, one group prototyped a closed-loop FPGA controller for a 20 kW bidirectional DC-DC converter including digital Proportional-Integral (PI) control and hardware-in-loop (HIL) testing. The FPGA managed the converter switches in real time, yielding a hardware-in-the-loop proof-of-concept for EV battery charge control. Another design used an FPGA to generate hysteresis-controlled PWM for both an inverter and a bidirectional charger, maintaining current THD below 5%. The result was that the FPGA-generated compensation currents effectively canceled grid harmonics while steering energy to or from the battery [3]. Wireless EV charging adds another power electronic dimension. Static inductive chargers require high-frequency inverter drive. FPGAs can produce sophisticated gate drive signals for wireless systems. An FPGA can implement real-time alignment correction algorithms (adjusting frequency or coupling) to maximize transfer efficiency. Meanwhile hybrid energy storage combining batteries with ultracaps can be coordinated as the FPGA balances power between slow (battery) and fast (capacitor) storages according to load dynamics. FPGA-based control of DC-DC and wireless converters yields highly integrated power electronics. Compared to separate controllers an FPGA can handle multiple converter channels and protection logic in one chip, which reduces component count. For example, one FPGA design integrated both inverter and bidirectional converter control in a single platform streamlining development. These implementations demonstrate that FPGAs can meet the rapid switching and feedback demands of EV-grade converters, while enabling advanced strategies (adaptive modulation, fault handling) beyond conventional fixed-function controllers.

Batteries ($LiFePO_4$, lead-acid) are used to store excess solar energy and provide a stable power supply compensating for the intermittent nature of renewable sources. They can also potentially feed energy back to the grid (V2G) during peak demand hours acting as an income source and enhancing grid stability [4]. WPT technology based on strongly coupled magnetic resonances is explored to eliminate the need for physical connectors simplifying the charging process and enabling dynamic charging while the vehicle is in motion. This significantly reduces charging times and minimizes congestion at charging stations [5]. The key functionalities of modern EV chargers include communication with the grid (IEEE 1547, open charge point protocol (OCPP), ISO 15118), demand response (DR) participation, and V2G power exchange. In a coordinated smart grid, EVs become flexible loads (or sources) that can smooth demand peaks and store excess renewable energy. For example, analyses show that intelligently managed EV charging can cut peak grid loads by 10%–15%, markedly improving resilience. Integrating EVs via demand side management (DSM) or DR programs allows shifting charging to off-peak times or responding to real-time prices. With robust DR, overall energy costs in a neighborhood can fall by 20% while grid reliability rises. The concept of V2G is extended so that when plugged in, an EV can supply energy back to the grid or local microgrid under control of the station. This bidirectional flow requires fast, fault-tolerant control logic. FPGA-based stations are well positioned to implement the necessary protocols and power conversion. In practice, an FPGA can handle high-speed communication (the ISO 15118 Plug & Charge handshake) and switch converters to reverse mode for V2G power flow. The station's FPGA could also execute real-time aggregation of V2G services dynamically bidding EV power into a frequency regulation market. Early prototypes (outside this abstract's specific literature) have shown that FPGA cores can meet the timing and security demands of V2G. The coupling of PV EV FPGA systems with the grid holds great promise. By coordinating charging and discharging, these stations act as distributed energy resources. They can absorb surplus PV power at midday then feed it back at peak times, leveling community demand. In this smart grid, FPGA-based controllers enable rapid decisions, reading meter/feeder measurements and controlling dozens of chargers in microseconds. This latency is crucial as the uncontrolled EV charging spikes can otherwise lead to transformer overloads and voltage dips. Thus FPGA-powered smart chargers become enablers of stable, optimized grids through DR and V2G.

Despite their advantages, FPGA-based charging systems face practical hurdles. FPGAs are complex to design as they require hardware description languages (HDLs, such as VHDL/Verilog) and expertise in digital design unlike more common microcontrollers. This can increase development time and may deter smaller manufacturers. Moreover, high-end FPGAs (with large logic and DSP blocks) can be costly, raising Bill of Material (BOM) costs of each charging station. Power efficiency of FPGAs is good for parallel tasks but static power (leakage) can be higher than an Application Specific Integrated Circuit (ASIC) or Microcontroller Unit (MCU). Another challenge is standardization as the electric vehicle supply equipment (EVSE) industry relies on protocols (International Electrotechnical Commission (IEC) 61851, ISO 15118) and certification (UL, CE) that focus on safety and interoperability. FPGA designs must still meet these standards, there is no open FPGA reference design for EVSE, so each vendor often creates proprietary solutions. This fragmentation can slow adoption of cutting-edge control. Finally, complex artificial intelligence (AI) on FPGA solutions require validation (proving that an RL scheduler is safe and reliable under all scenarios which is an active research area).

Future work will push further integration, optimization, and standardization. On the hardware front we expect more SoC FPGAs (combining Advanced RISC Machine [ARM] cores and FPGA fabric) to be used, simplifying co-design of software and hardware for charging stations. Power-wise,

GaN and SiC converters may be tied to FPGA controllers for even higher efficiency and switching speeds [6]. In AI and control, deeper neural network accelerators on FPGA could enable predictive energy management (forecasting PV output and EV arrivals). Federated learning across stations could coordinate city-wide DR. On the grid side V2G could evolve to full Vehicle to Home (V2H) or Vehicle to Grid to Vehicle (V2G2V) cycles with FPGAs orchestrating aggregate battery capacity for renewable smoothing. This will translate into networks of stations sharing data and load information managed by regional energy management centers. Renewable integration could also expand by adding wind turbines or grid batteries to the station, which would simply be another FPGA-controlled resource. Sustainability considerations (circular electronics, low-carbon fabs) may steer future FPGA choices but the reprogrammability ensures that stations can adapt to new norms (future EV standards or grid codes) without hardware overhaul. The convergence of solar generation, EV charging, and FPGA intelligence is a rich research area. By combining high-speed hardware control with renewable energy and smart algorithms these systems aim to be scalable, sustainable, and optimally managed. Continued innovation in power electronics, FPGA architectures, and energy optimization algorithms will further unlock the potential of solar-powered, FPGA-managed EV charging infrastructure [3]. The integration of solar PV systems with FPGA-based real-time control represents a transformative step toward intelligent, sustainable, and resilient EV charging infrastructure. By leveraging high-speed, reconfigurable hardware, these systems can dynamically manage energy flows, enhance power quality, and support advanced grid services such as DR and V2G participation. Despite challenges related to design complexity and standardization, continued innovation in FPGA architectures, power electronics, and AI-driven control strategies holds great promise. Ultimately, the convergence of renewable energy, advanced power management, and smart grid connectivity positions solar-powered, FPGA-managed charging stations as a cornerstone of future clean mobility ecosystems.

In the future, combining solar energy, advanced electronics, and FPGA-based control will help make EV charging stations even smarter and more useful. There are still some problems to solve, like making FPGAs easier to use, which makes sure that systems are affordable and agreeing on how everything should work together. But with new ideas and constant improvements these charging stations will become more reliable and common. As a result solar-powered, smart charging stations run by FPGAs will likely become a key part of a cleaner and greener way for people to travel.

6.2 FPGAS: THE SMART BUILDING SYSTEM'S DIGITAL BRAIN

As the name suggests, a FPGA is an integrated circuit that can be configured by the final user after it is manufactured. In fields like wireless telecommunications, robotics, digital signal processing, image recognition, increasingly renewable energy systems, and smart building infrastructure, FPGAs are essential for building control platforms for real-time applications. In intelligent building particularly solar-powered buildings and EV charging stations, FPGAs provide the mind for adaptive, high-speed control. Their reconfigurability enables them to rapidly process sensor inputs, switch energy sources, balance loads, and apply custom control algorithms without using fixed hardware or sequential. An FPGA is programmed with an HDL like Verilog or VHDL so that the designer can specify the logic and the behavior of the system at a low level. Internally, an FPGA holds between thousands and millions of programmable logic blocks, which may be interconnected through a hierarchy of reconfigurable interconnects—basically creating a connection of logic gates that can be wired in almost any topology. For solar-driven applications, control algorithms are executed on the FPGA to achieve activities like MPPT, load prioritization, and EV charging coordination. Standard hardware implementations would include special control signals like CHIP_SELECT (input) and DRIVER_IN (output), which can be linked to physical indicators like LEDs or any other devices for the display of visual feedback regarding power routing or logic execution [6].

6.2.1 Basics of a FPGA Design

An FPGA consists of the following components (refer Figure 6.1):

1. **Configurable Logic Blocks (CLBs)**: The heart of an FPGA. A CLB contains a combination of flip-flops, and other digital circuits for performing combinational and sequential logic.
2. **Interconnects**: A programmable network of interconnects that connects various digital logic blocks such that the designer may design complex circuits.
3. **I/O Blocks**: Link the FPGA to external peripherals such as sensors, analog-to-digital converters (ADCs), DACs, memory, or communication buses (UART, SPI, I2C).
4. **Clock Management Units (CMUs)**: Use for clock signal control and routing within the chip.
5. **Memory Blocks**: FPGAs normally have block RAMs (BRAMs) for storing data in between or to use for buffers and queues.

Xilinx Zynq-7000 series (Zed Board/ZC702/ZCU104):

- **FPGA Fabric**: Artix-7 or Kintex-7
- **CPU Core**: Dual-core ARM Cortex-A9 (PS—Processing System)

Ideal for: Complex control systems (MPPT, BMS), AI integration, mixed signal processing.

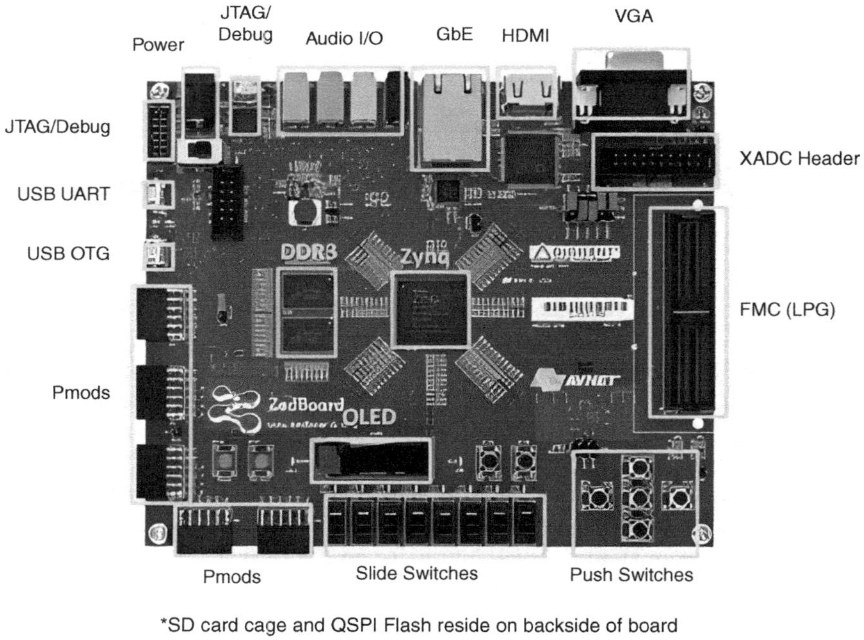

FIGURE 6.1 Representation of Xilinx Zynq-7000 series FPGA board.

Advantages:

- **Hybrid**: CPU for OS and FPGA for real-time control.
- HDMI, Ethernet, USB, and GPIOs.

Use Case: Widely used in renewable energy, smart grid, and EV charger prototypes [7].

6.2.2 FPGA Design Flow for Smart Energy Applications

The FPGA technology–based design and development, e.g., solar-integrated EV charging systems, involves an iterative and systematic process. Figure 6.2 shows the typical FPGA design process, which starts from initial design entry and ends with real-time in-circuit verification on physical hardware (Figure 6.2).

6.2.2.1 *Source design documentation*

This is the structural block wherein the behavior of the system is modeled using HDLs such as Verilog or VHDL. The program defines the FPGA behavior in control, decision-making, signal processing, and other functionality in terms of input conditions such as solar power availability or battery charge.

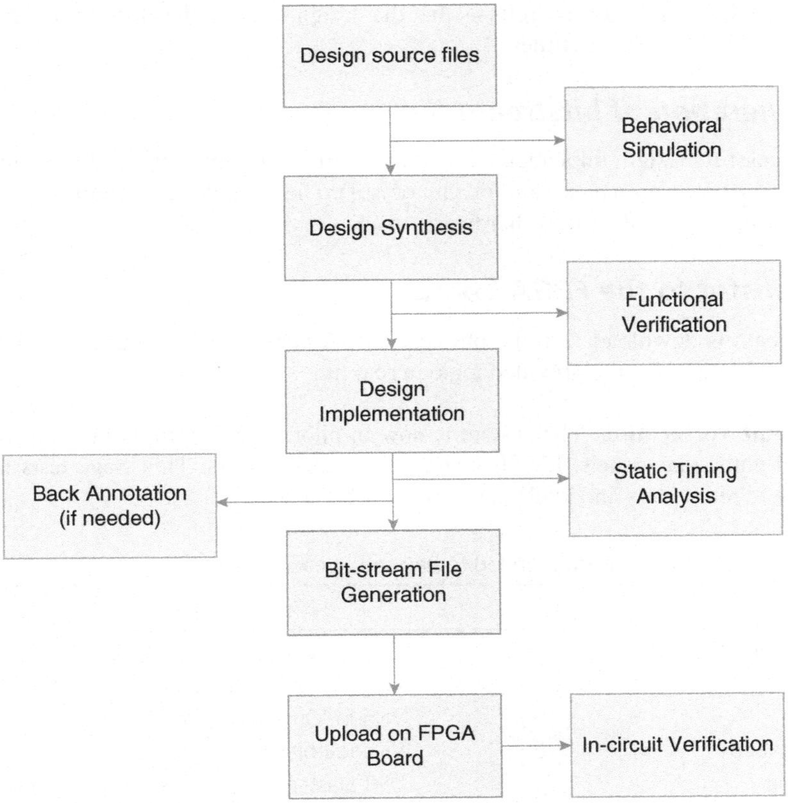

FIGURE 6.2 Field-programmable gate arrays (FPGAs) design flow.

- **Behavioral Simulation**: By doing so, the HDL code runs in an environment that guarantees the logical correctness of the code regardless of target hardware that it is going to be implemented for. This is beneficial in terms of bug catching during programming and checking functionality at a conceptual level.

6.2.2.2 *Synthesis*

Synthesis converts the HDL description into netlist and low-level gate-level description of the design in terms of target FPGA-supported primitives. Mapping is from abstract to physical FPGA primitives such as Look Up Tables (LUTs), flip-flops, and multiplexers.

- **Functional Verification**: The code defines the functional behavior of the FPGA in terms of control, decision-making, signal processing, and other tasks based on input parameters such as the presence of solar power or the status of battery charge.

6.2.2.3 *Design implementation*

Implementation consists of three sub-steps:

a. **Placement**: Places logic blocks into certain physical locations on the FPGA.
b. **Routing**: Connecting them to interconnect resources.
c. **Optimization**: Reconfiguring placement and routing to satisfy performance requirements.
   - **Static Timing Analysis**: Tells us that the design satisfies the close timing requirements such as hold and setup times.

6.2.2.4 *Generation of bitstream file*

Following successful design implementation and verification, the FPGA development tool (e.g., Quartus or Vivado) produces a bitstream file (.bit or .sof). The file holds the required configuration data employed for programming the FPGA hardware.

6.2.2.5 *Transfer to the FPGA board*

Then the bitstream is downloaded to the physical FPGA board via JTAG programmer or USB. The FPGA is prepared to execute the provided logic in real time.

- **In-Circuit Verification**: The system is now in pilot operation in its target environment—coupled with solar panels, EV charging points, and sensors. This stage tests the system's reaction to real inputs and loads the system (Table 6.1).

TABLE 6.1 Summary of tools recommended to be used at each step

STEP	*TOOL EXAMPLE*
HDL design and simulation	Vivado, ModelSim, QuestaSim
Synthesis	Xilinx Vivado, Intel Quartus
Implementation	Vivado, Quartus (Place & Route)
Bitstream generation	Vivado (.bit), Quartus (.sof)
Hardware upload	USB Blaster, JTAG, Xilinx Platform Cable
In-circuit testing	Oscilloscopes, logic analyzers, sensor interfaces

6.2.3 System Architecture: FPGA Implementation in Solar-Powered EV Charging

We will explain how the FPGA acts as the central controller, which interacts with the different elements of the hardware and software system of the smart building. On a high-tech building system with solar-powered EV charging, the FPGA is the master component and control device, coordinating real-time energy distribution among PV sources, battery storage units, EVs, and the power grid. System architecture overall is optimized to deliver maximum energy efficiency, minimized grid dependency, and increased reliability and security of energy supply. The core of this design is the FPGA, which manages multiple subsystems simultaneously because it has the capability to process in parallel and has low-latency response time [8,9].

6.2.3.1 The key factors in the architectural model

1. PV array
   - Converts sunlight into DC electricity.
   - It is connected to the FPGA-based MPPT module to provide maximum power extraction disregarding varying brightness [8].
2. DC-DC converter
   - Controls the voltage of the PV panels to the requirements of the battery or EV.
   - It will be controlled with the FPGA by using a high-frequency PWM signal for precise voltage regulation [10].
3. Battery energy storage system (BESS)
   - Stores excess energy when too much solar energy is generated for use during times of low sunlight.
   - Regulated by an FPGA-based BMS that monitors SoC, temperature, voltage, and current [11].
4. EV interface/charging station (EVSE)
   - System output end where the vehicle plugs in order to charge.
   - The FPGA controls the charging current based on available solar power, battery status, and user preference [12].
5. Grid interface
   - Offers backup to grid electricity.
   - Enables V2G energy transfer under FPGA control [13].
6. Sensor network
   - Current sensors, voltage sensors, and temperature sensors.
   - Sensor data are captured with ADC and it will be processed in real time by the FPGA [14].
7. Control interface (user/building management system)
   - A display or wireless interface indicates system status (e.g., power source, charging level).
   - The communication with the Building Management System can be established through protocols like UART, SPI, I2C, or higher-level protocols like Modbus or MQTT [9].

6.2.3.2 Primary function of the FPGA

The FPGA connects all critical system components to create a tightly coupled energy management loop:

- Performs MPPT algorithms (P&O, Incremental Conductance) for solar harvesting optimization [8].

- Takes multiple decisions to redistribute power to the EV or to the battery or to the grid, depending on real-time sensing [12].
- It tracks real-time voltage, current, and thermal limits to protect hardware and long battery/EV life [11].
- Generates prompt and responsive PWM signals to control the switching mechanisms in DC-DC converters or inverters [10].

This fine-grained and responsive control is critical for those energy-sensitive applications, such as charging EVs, where power levels are potentially high and in flux, and where power sources, like the sun, are variable and intermittent [14].

6.2.3.3 Conventional energy transfer pathways

- **Daytime (Full Sunlight)**: PV → FPGA → MPPT → DC-DC → EV (charge directly), or PV → BESS (storage)
- **Partial Sunlight or Cloudy**: PV + BESS → EV (FPGA dynamically balances the load)
- **Nighttime or Low Energy**: BESS → EV, or Grid → EV (failover)
- **Bidirectional (V2G Enabled)**: EV → Grid at peak time (if battery is surplus)

6.2.3.4 Benefits of this architecture

- **Energy Efficiency**: Straightforward solar-to-EV or battery direct route without waste conversion loss.
- **Flexibility**: Simple to accommodate energy policy changes (e.g., DR, time-of-use tariffs).
- **Reliability**: Protection of hardware using FPGA-driven fault detection (e.g., overcurrent, thermal runaway).
- **Sustainability**: Reduces carbon footprint by optimal use of renewable resources.

6.2.4 Principal FPGA Interfacing Modules

The intelligence of an FPGA-based power system is derived from its mode of interaction with its peripheral devices—EVs, batteries, grid systems, converters, and sensors. The next section discusses the major functional blocks that are implemented inside the FPGA and how they enable smart, real-time control in a solar-powered EV charging system.

6.2.4.1 MPPT controller module

The MPPT controller is responsible for adjusting the PV array's operating point to obtain the maximum power available at any given moment. Due to real-time environmental changes like temperature or irradiance, MPPT algorithms based on FPGA, such as Perturb and Observe (P&O) or Incremental Conductance, offer high convergence speed with zero energy loss [8]. FPGAs are better than the normal microcontrollers in this regard as they are able to sample and control continuously with almost zero delay. Thus, the power consumed from the solar array is greatly optimized [9].

6.2.4.2 PWM generator for power converters

The FPGA generates high-frequency PWM signals to power DC-DC converters (buck, boost, or bidirectional) to control the voltage between the EV charging circuits, solar panels, and batteries. This timing accuracy ensures clean switching, low harmonic distortion, and greater efficiency—especially if SiC or GaN-based power devices are employed in the design [12].

6.2.4.3 Sensor interface module

The sensor network feeds the FPGA the required real-time data—i.e., voltage, current, temperature, and irradiance. Depending upon the complexity of the system, these sensor signals may be interfaced directly via the FPGA's onboard ADCs (if present) and connected via external ADCs via SPI, I2C, or UART protocols.

6.2.4.4 EVSE communication interface

The FPGA talks to the EVSE, either direct power flow control with converters and relays or coordinating communications through EV protocols (CAN, UART) as needed. It provides safe charging start, real-time monitoring of charging current, priority-based charging support optional, and bidirectional V2G communication if installed [13].

6.2.4.4.1 Use-case scenario examples in FPGA-based solar EV charging systems

In order to grasp the working relevance of an FPGA in a smart energy network, you must understand real working conditions. The following examples describe how an FPGA controls different sources of energy (solar, battery, and grid), controls real-time channeling of energy, and reacts interactively to dynamic conditions in a solar-integrated EV charger system.

Use Case 1: Daytime operations with full solar availability

Scenario: Clear day with high irradiance and low grid dependency.

System Behavior:

- The FPGA is supplied with high solar voltage and current from the sensors.
- MPPT module regulates the operating point in real time for maximum power.
- Buck converter is controlled by PWM pulses to provide the regulated voltage to the EV charger.
- The excess solar power is stored in the battery by charging.
- System displays "Charging from Solar" on LEDs or display interface:

  Benefit: Zero grid draw, maximum possible use of renewables, and simultaneous EV charging + storage.

Use Case 2: Variable solar power on cloudy days

Scenario: Periodic cloud cover decreases and varies solar radiation.

System Behavior:

- FPGA senses the reduction of solar current from sensor input.
- The MPPT controller continues to adjust based on continuously changing conditions.
- The energy routing logic checks if to recharge the EV repeatedly in partial solar + battery backup.

  Temporarily stop charging or go to grid power if needed.
- Live adjustment of the charging current prevents power loss.

  Benefit: Seamless transition without the intervention of the user, providing seamless EV charging with dynamic power sharing.

Use Case 3: Nighttime operation (no solar)

Context: No sunlight to draw on once night falls.

System: Behavior

- The FPGA registers zero output from the PV array.
- It confirms the battery's SoC:
  - If sufficient, battery powers the EV charger.
  - If low, the FPGA switches to the grid as a fallback.
- Real-time feedback is used to prevent over-discharging the battery.

Benefit: Long-term nighttime EV charging with smart source scheduling.

Use Case 4: Prioritized Smart Charging

Scenario: They do have some EVs charged in, but very little solar/battery input.

System Behavior:

- FPGA is given EV priority levels or arrival timestamps.
- Scheduling algorithm (rule-based or Machine Learning [ML]-based) determines:
  - Which vehicle charges first.
  - Each EV's comparative charging rate.
  - Ideal current distribution to minimize waiting time.

Benefit: Optimal, fair utilization of energy sources to the user's satisfaction.

6.2.5 Finite State Machine

For intelligent control, an effective FPGA-based EV charging system is dependent upon a robust Finite State Machine (FSM) for switching among energy modes like solar charging, battery backup, idle mode, and error recovery. The FSM provides low-latency, deterministic, and real-time response system state-based switching such as fault detection, PV availability, EV plug-in status, and battery SoC. FSM states (SO–S10) are designed in a way to handle power routing logic, fault management, EV charging decision, and MPPT control logic (refer Figure 6.3 and Tables 6.2 and 6.3).

6.2.6 Example FPGA HDL Block Mapping

Each state can be mapped to a separate module or process block in VHDL/Verilog (Figure 6.4).

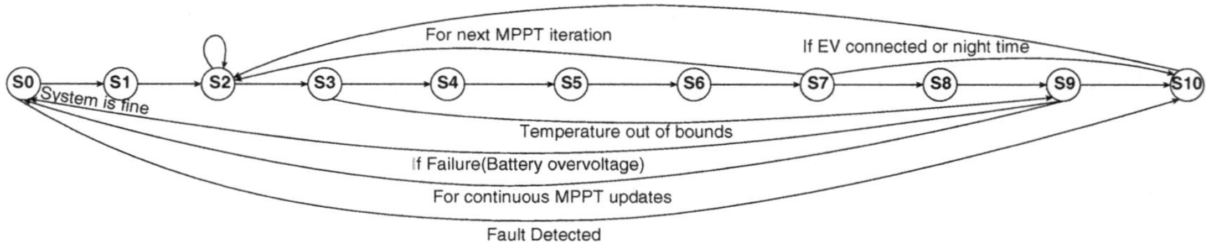

FIGURE 6.3 State diagram for solar charging EV using FPGA.

TABLE 6.2 Representation of state diagram for solar

STATE	*NAME*	*FUNCTION*
S0	System initialization	Loads configuration, initializes FPGA registers, checks PV and EV interfaces. Transitions to S1 if OK; else S10 (Idle/Error).
S1	Resource configuration	Fetches PV capacity, battery size, user charging preferences, and building load requirements.
S2	MPPT operation	Executes the MPPT algorithm to extract max power from PV. Continuously loops here during sunlight availability.
S3	Battery monitoring	Reads voltage, SoC, and temperature from the BMS. If normal, proceeds to S4; if fault (e.g., overvoltage/thermal), jumps to S10.
S4	Power flow control	Manages charging or discharging through DC-DC converters. Handles energy routing between PV, battery, and EV.
S5	Load balancing decision	Determines power allocation between EV, building load, and battery. Evaluates grid support if required.
S6	EV scheduling and priority control	Selects which EV to charge (if multiple) based on SoC, departure time, and priority (fast/normal).
S7	Data logging and monitoring	Logs energy statistics, MPPT efficiency, charge time, and system events.
S8	Simulation and testing mode	Used during HIL (Hardware-in-the-loop) development or offline testing.
S9	Error handler	Catches MPPT faults, sensor disconnects, or unsafe charging events. Resets to S0 or remains in safe loop.
S10	Idle/sleep mode	Entered during night (no solar), no EV connected, or fault state. Wakes up on trigger from PV (S2) or user input (S3).

TABLE 6.3 Representation of FSM transitions

CONDITION	*TRANSITION*
System initialized correctly	S0 → S1
Solar available	S1 → S2
If EV connected/nighttime	S7 → S10
MPPT stable and working	S2 → S3
Battery safe zone	S3 → S4
Power demand exists	S4 → S5
Multiple EVs/load imbalance	S5 → S6
EV selected	S6 → S7
Data logged and charging complete	S7 → S8 → S10
Fault detected (e.g., overvoltage, thermal)	Any → S9 or S10
Recovery or wake-up trigger	S10 → S2 or S3

6.2.7 Simulation Output

Inputs set (Figure 6.5):

- ev_connected = 1
- solar_available = 1
- battery_safe = 1

```
module  fsmovdesign  (
      input    clk,
      input    reset,
      input    av_connected,
      input    solar_available,
      input    battery_safe,
      input    error_detectod,
      Output  reg [3:0] state
);
Parameter   S0_INIT            = 4'd0;
Parameter   S1_CONFIG          = 4'd1;
Parameter   S2_MPPT            = 4'd2;
Parameter   S3_BATT_CHECK      = 4'd3;
Parameter   S4_PWR_FLOW        = 4'd4;
Parameter   S5_LOAD_BAL        = 4'd5;
Parameter   S6_SCHED           = 4'd6;
Parameter   S7_DISPLAY         = 4'd7;
Parameter   S8_LOGGING         = 4'd8;
Parameter   S10_Error          = 4'd10;
Parameter   S11_IDLE           = 4'd11;
reg  [3:0]  next_state:

always @(posedge  clk  or  posedge  reset) begin
      if  (reset)
          state  <=  S0_INIT;
      else
          state  <=  next_state;
end

always @(*) begin
       case  (state)
            S0_INIT:  begin
                 next_state = S1_CONFIG;
            end
            S1_CONFIG: begin
                 next_state = Solar_available  ?  S2_MPPT  :  S11_IDLE;
            end
            S2_MPPT:  begin
                 next_state = battery_safe  ?  S3_BATT_CHECK  :  S10_ERROR;
            end
            S3_BATT_CHECK: begin
                 next_state = ev_connected  ?  S4_PWR_FLOW  :  S11_IDLE;
            end
            S4_PWR_FLOW: begin
                 next_state = S5_LOAD_BAL;
            end
            S5_LOAD_BAL:  begin
                 next_state = S6_SCHED;
            end
            S6_SCHED: begin
                 next_state = S7_DISPLAY;
            end
            S7_DISPLAY: begin
                 next_state = S8_LOGGING;
            end
            S8_LOGGING:  begin
                 next_state = S2_MPPT;
            end
            S10_ERROR: begin
                 next_state = S0_INIT;
            end
            S11_IDLE:  begin
                 next_state  =  solar_available ? S2_MMPT  :  S11_IDLE;
            end
            default:  begin
                 next_state = s0_INIT;
            end
       endcase
end
endmodule
```

FIGURE 6.4 Verilog code for FSM.

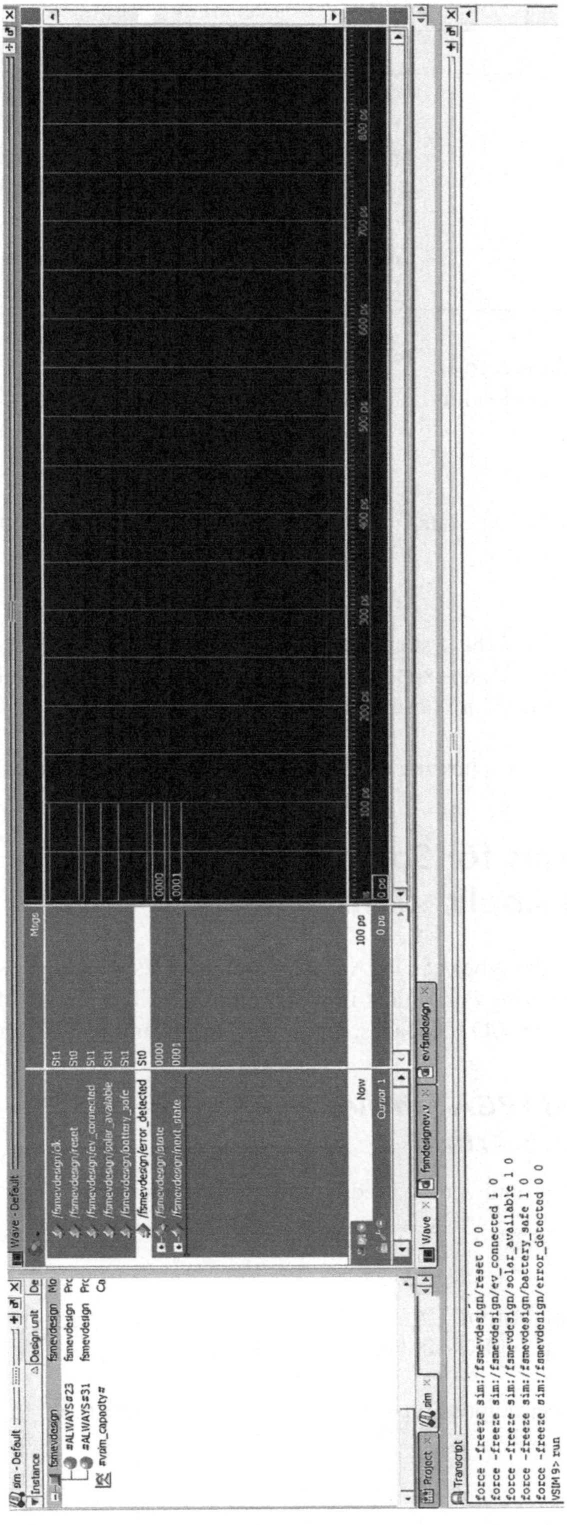

FIGURE 6.5 Output waveform for Verilog code.

TABLE 6.4 Representation of selection criteria for FPGA in smart EV charging

FEATURE	*IMPORTANCE*
Low power consumption	Must operate reliably in solar-powered systems
Number of I/Os	Needed for sensors, drivers, comms, PWM
Onboard ADCs or SPI/I²C support	For sensor interfacing (current, voltage, temp)
PWM capability	Control of buck/boost converters and inverters
Soft processor support	For hybrid control (e.g., MicroBlaze/Nios II)
Reconfigurability	Allows future upgrades (new MPPT algorithms, EV modes)
Communication support	UART, CAN, Ethernet for smart building integration

- reset = 0 (meaning FSM is active)
- error_detected = 0 (system healthy)

State transition:

- state moved from 0000 → next_state = 0001 (from St0 to St1), indicating proper FSM behavior.

Interpretation:

- The FSM transitioned out of the system initialization state (S0) to the resource configuration or PV/battery check state (S1) as expected when all preconditions were met.
- No glitches or floating values are observed—signal stability is confirmed.

The system did not enter error state, showing that fault logic is functional and clean.

6.2.8 FPGA Platforms for Solar-Based EV Charging Applications

In this section, we will explore the types of FPGA boards and families that are most suitable for implementing solar-powered EV charging controllers in smart buildings. The choice of FPGA depends on performance, power consumption, I/O interfaces, reconfigurability, and cost (Table 6.4).

6.2.8.1 Recommended FPGA families and boards—Xilinx Spartan-6/Artix-7

Why? Cost-effective, low-power, and widely used in power electronics.
Features:

- 200–500 user I/Os
- DSP slices for MPPT algorithm
- Artix-7 supports high-speed I/Os and more logic blocks

Application Fit: Ideal for academic/prototype solar EV chargers (Figure 6.6).

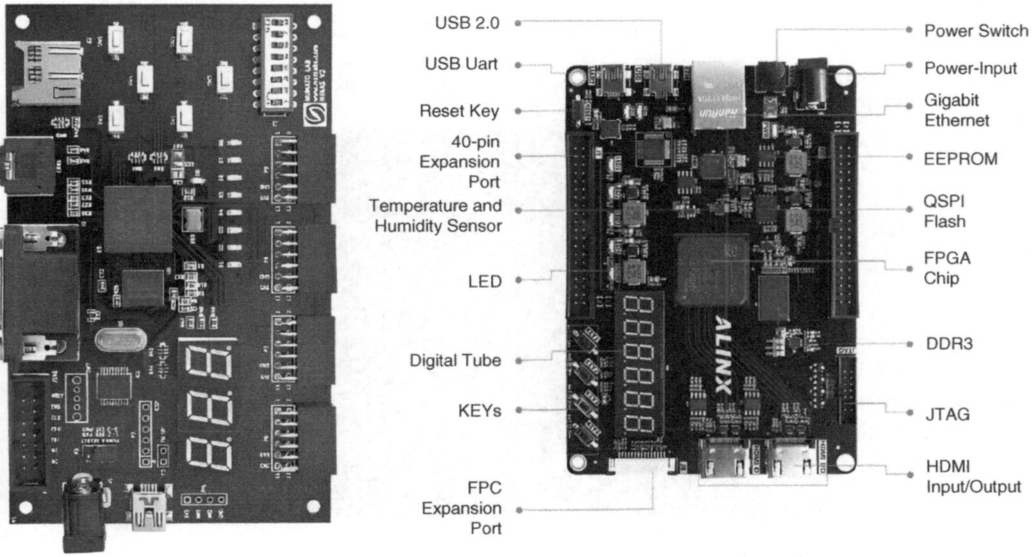

(a) (b)

FIGURE 6.6 (a) Representation of Artix-7 board and (b) representation of Xilinx Spartan-6.

6.2.8.2 Intel (Altera) Cyclone IV/V

Why? Balanced logic + power + soft-core integration (Figures 6.7 and 6.8).
Features:

- Nios II processor support
- PLLs, ADC-ready interfaces
- **Cyclone V**: includes SoC variants with ARM cores

Boards:

- DE0-Nano (Cyclone IV)
- DE1-SoC (Cyclone V with ARM Cortex-A9)

Application Fit: Useful for building BMS + MPPT + UI in one platform.

6.2.8.3 Xilinx Zynq-7000 series

Why? Combines FPGA fabric with a dual-core ARM processor.
Features:

- Best for hybrid control + networking
- Full Linux/RTOS support
- Built-in ADC interface and secure boot

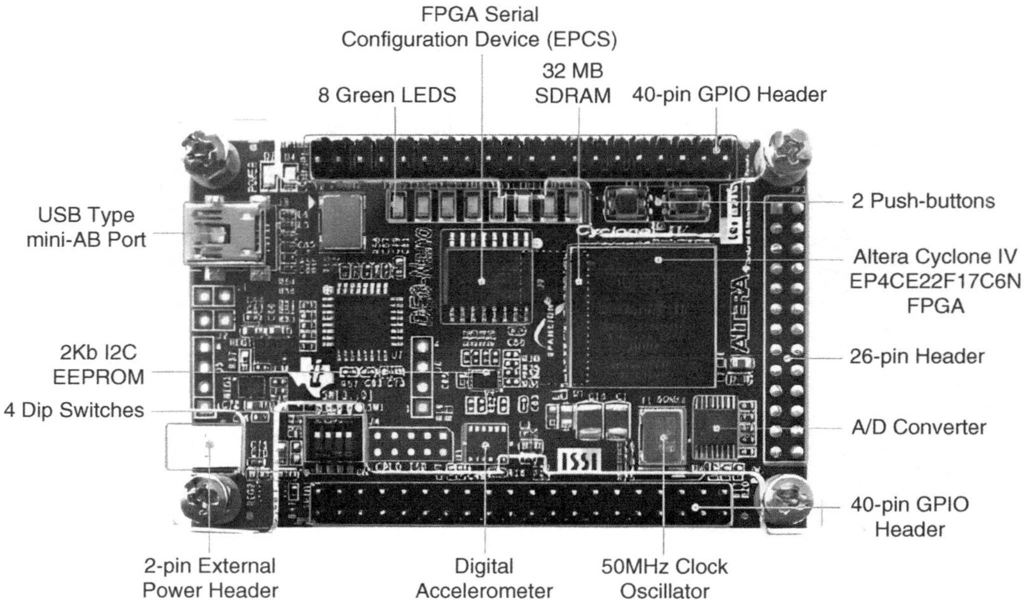

FIGURE 6.7 Representation of DE0-nano (Cyclone IV) board.

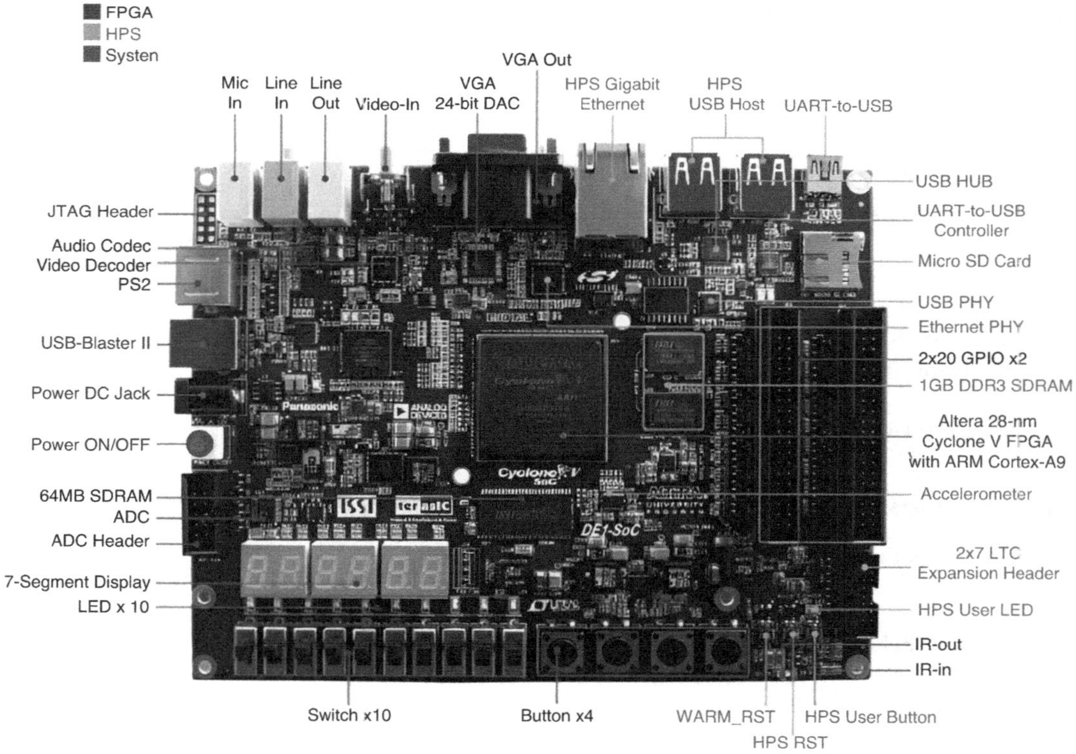

FIGURE 6.8 Representation of DE1-SoC (Cyclone V with ARM Cortex-A9).

Boards:

- ZedBoard
- Zybo Z7

Application Fit: High-end smart building applications with cloud support, dashboards, and ML (Figures 6.9 and 6.10).

6.2.8.4 Lattice iCE40/MachXO2

Why? Ultra-low power and compact size.
Features:

- Targeted for edge computing
- Can implement simple control FSM, PWM
- Less I/O and logic but superpower-efficient

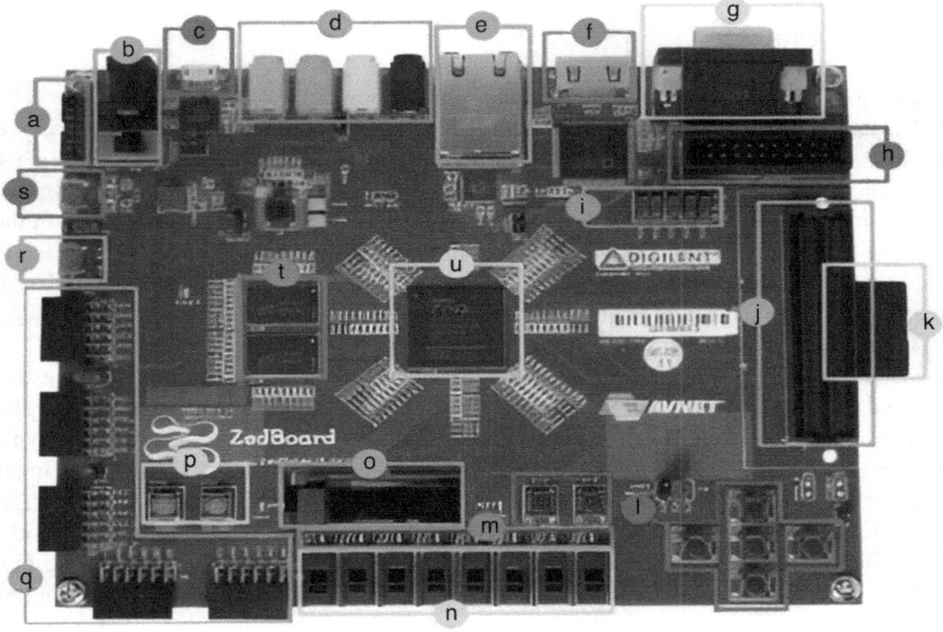

FIGURE 6.9 Representation of Zed board.

FIGURE 6.10 Representation of Zybo Z7 board.

Boards:

- iCEstick, iCEBreaker

Application Fit: Minimalist or portable solar charging stations.

FPGAs have also turned out to be very versatile and efficient platforms for implementing real-time control and intelligent energy management in smart buildings. In solar-powered EV charging systems, the FPGAs provide an interesting blend of speed, programmability, and parallel processing that makes them most suitable for embedded decision-making where timing and responsiveness are paramount. At this point, we have seen how the FSM-based control logic, designed using Verilog HDL, can efficiently handle state transitions for solar availability, battery protection, MPPT tracking, and EV charging connection events. FPGAs' programmable nature enables such FSMs to be reprogrammed, debugged, and updated with control tactics changes.

Besides, the MPPT algorithm, being a fundamental block toward solar energy harvesting maximization, was also successfully employed within the FPGA substrate. The results of the simulation showed that not only did the system monitor real-time voltage and current data but also dynamically regulated the duty cycle of the power converter for optimal charging efficiency. By comparing different FPGA platforms, we also described how the designer will be able to opt for boards such as lightweight boards, Lattice iCE40 for low-power systems, or Zynq SoC boards for edge intelligence and connectivity. The selection should be based on system complexity, interfaces needed, and budget. In essence, FPGAs bridge the gap between hardware precision and software programmability, enabling smart buildings to:

- Govern renewable energy autonomously,
- Efficiently schedule EV charging, and
- Respond quickly to loading and environmental conditions.

Their hardware-level performance, reprogrammability, and deterministic control guarantee FPGAs to remain a central component in the roll-out of energy-aware smart infrastructure in the next few years.

6.3 SOLAR CHARGING USING ON-GRID TECHNIQUES FOR EV CHARGING

The transport industry is in the process of a paradigm shift with increasing use of EVs. The imperative spurs the move to mitigate greenhouse gas emissions, enhance air quality, and decrease reliance on fossil fuels. The shift presents new challenges, which include energy supply and grid management. EV charging stations must be built at a fast pace and incorporated alongside renewable energy supplies to be sustainable. Solar PV systems have also become a clean, ubiquitous, and scalable source of energy that can be utilized to charge EVs in combination with on-grid systems. Solar EV charging can supply predictable power while enabling the support of grid operation through functions such as peak shaving and V2G capabilities [15] (Figure 6.11).

An on-grid solar system is also referred to as a grid-interactive or grid-tied solar system. It is a PV system connected directly to the public electricity grid. This implies that your solar panels provide electricity, and this electricity can either be utilized by your household, or excess electricity can be supplied back into the grid. On the other hand, if your solar panels are not generating sufficient power, your household can be provided with electricity directly from the grid [16].

6.3.1 Electrical Characteristics of a Solar Panel

We all know that there are two types of electricity, one is AC, while the other is DC. The power produced by solar panels is of the second type, namely DC power, similar to that of batteries. In general, the open-circuit voltage (VOC) produced by the solar panel is around 0.5 or 0.6 V, but this voltage depends on the load demand.

The power produced from the solar panel is proportional to the solar irradiance at the surface of the panel.

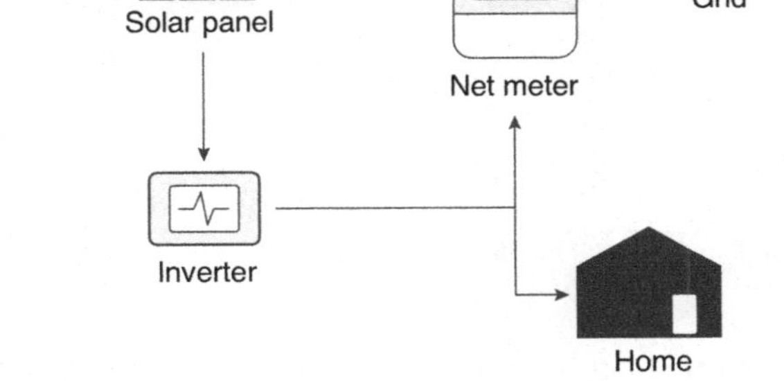

FIGURE 6.11 Diagram of an on-grid solar system. It shows the normal configuration and energy flow of an on-grid solar photovoltaic (PV) system.

But there is a limit to the maximum power generated from a single solar cell, no matter how high the solar irradiance is. This is known as the maximum power, denoted as P_{max}. The efficiency of converting the sunlight hitting the solar panel depends on the type of semiconductor material that the cell is made from. The most common types of semiconductor materials used for the manufacture of PV panels are silicon, gallium arsenide, cadmium sulfide, etc.

Commercially available PV solar cells have power ratings that help us decide on the type of converters that are needed, and the total number of panels that will be required to generate the required power. The maximum power that will be generated from the solar panel is mathematically denoted as:

$$P_{max} = V_{out} \times I_{max}$$

where P_{max} is the maximum deliverable solar power in watts.

- V_{out} is the output voltage from the solar cell in volts.
- I_{out} is the maximum cell current in amperes.

In Figure 6.12a, the *I-V* curves illustrate how the output current of the PV array varies with voltage across different temperatures:

- The curve begins at the short-circuit current (ISC) on the y-axis. At this point, the voltage is zero, but the current output is at its maximum.

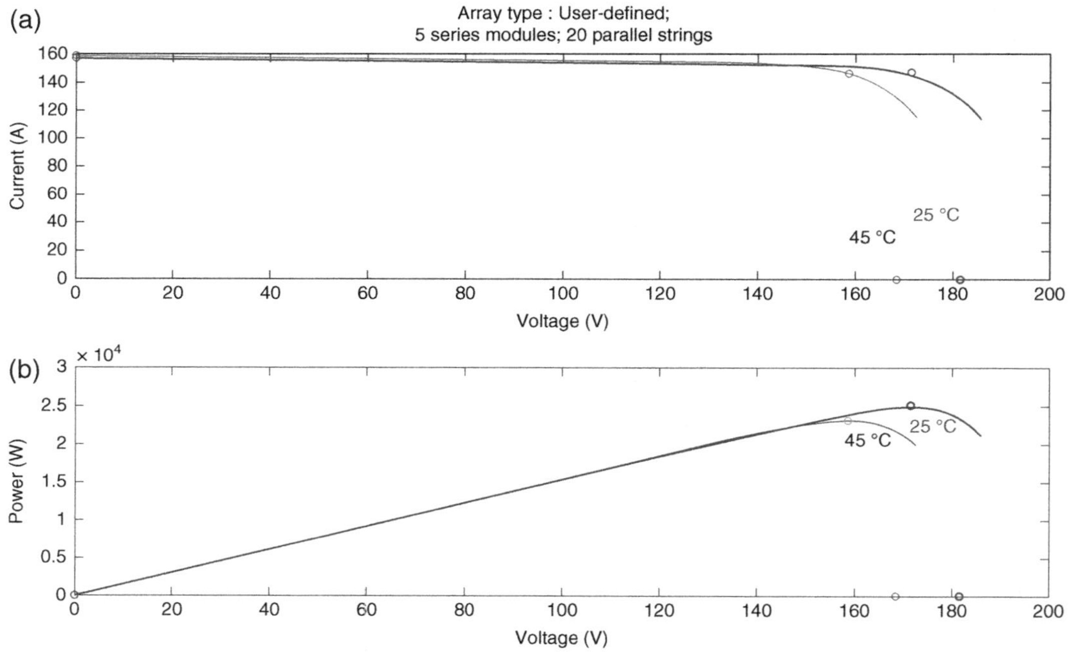

FIGURE 6.12 (a) A graph plotted between the current and the output voltage of the solar panel and (b) a graph plotted between the output power and output voltage for different temperatures.

- The current stays almost constant as the voltage rises until it reaches a critical point where it starts to drop sharply. The array is getting close to its maximum power point (MPP) at this inflection point.
- After this critical point, the current drops off rapidly as the voltage rises until the curve reaches the VOC, at which point the current drops to zero.
- The two temperature curves demonstrate that temperature significantly impacts the PV output. The entire curve shifts to the left at higher temperatures because of a lower VOC and a slight decrease in current. The inherent negative temperature coefficient of voltage in PV modules is aptly demonstrated by this phenomenon.

In Figure 6.12b, in the bottom subplot (referring to a typical representation of *P-V* curves), the *P-V* curves show how the electrical power output of the array varies with voltage:

- Initially, power increases with voltage as both voltage and current contribute positively to the overall output.
- The curve subsequently peaks at a definite point, which is the MPP. Here, the PV system is most efficient and provides its highest possible power output.
- Following this peak, power starts to fall. This is because the steep drop in current, as voltage keeps increasing, makes up for any incremental increase in voltage.
- The peak of each curve represents the optimum operating point of the PV system for that given temperature and irradiance level. The fact that a shift in the MPP toward lower voltages is seen at elevated temperatures highlights the importance of adjusting the operating point dynamically. It is exactly this dynamic adjustment that MPPT algorithms like P&O are meant to achieve.

6.3.2 Energy Flow in On-Grid Solar Charging System

Energy flow in on-grid solar charging system:

1. **Solar Energy Collection**: It starts with the Sun, which sends solar energy to the solar panels. The panels, made of PV cells, turn the sunlight directly into DC electricity.
2. **DC to AC Conversion**: The DC electricity produced by the solar panels goes into an inverter. The inverter's main job is to change this DC power into AC power, which is the usual type of electricity used in homes and works with the national electricity grid.
3. **Supplying the Home and Engaging with the Grid**: From the inverter, AC electricity goes to the house. The electricity produced by the solar system initially serves the immediate electricity needs of the home appliances. Any remaining electricity produced that is not used by the home is exported to the main grid.
4. **Net Metering**: The Net Meter is the key to tracking this energy transaction. It measures precisely the electricity taken from the grid by the house and the excess electricity fed into the grid from the solar system. Two-way metering permits "net metering" billing, in which customers are usually credited with the electricity they feed back to the utility.
5. **Grid Connection**: The whole system is connected to the grid with no break, so the home can use power when solar output is low and supply power when output is high, maximizing usage of energy and even cutting down electricity bills.

This architecture supports multiple modes of operation:

1. **PV-to-EV**: Direct solar charging during the day.
2. **PV Grid to EV**: When solar is insufficient, grid power supplements the load.
3. **V2G**: Stored energy in EVs is discharged back to the grid to support demand.

Given the intermittent nature of solar power generation and the time-varying patterns of EV charging, the combination of solar power and EV charging equipment presents complex energy management challenges. Generation-load mismatches can result in component overloading, energy waste, and issues with power quality if they are not properly controlled. Additionally, precise synchronization and coordination are required for two-way energy streams in grid-connected systems, such as energy being extracted and injected into the grid. Control strategies are the key to handling this dynamic relationship. They improve power harvesting from renewable energy sources, optimize energy distribution, ensure grid compatibility, and use battery storage devices. As solar EV charging systems become more common and complex, integrating smart control systems is necessary for safety, reliability, and efficiency.

Successful power management is paramount to the performance and dependability of on-grid solar EV charging systems. Some control strategies are utilized:

1. **Maximum Power Point Tracking**: MPPT algorithms are utilized to obtain the maximum power available from PV panels. Methods are P&O, incremental conductance, and fuzzy logic controllers. Fuzzy logic controllers work best under changing environmental conditions because they can manage nonlinearities as well as uncertainty without requiring an accurate mathematical model.
2. **SoC Management**: Battery SoC monitoring guarantees battery lifetime and safety for energy storage systems. Charging and discharging are controlled to avoid overcharging or deep discharge, maximizing battery health.
3. **Load Forecasting and Scheduling**: Sophisticated Energy Management Systems (EMSs) apply historical data and machine learning to forecast load demand and solar generation. These forecasts improve the charging schedule and make solar a priority. In our case study, an FPGA is utilized for both dynamic load balancing and MPPT tracking. This technique guarantees that load demands are met effectively without compromising system stability and enhances real-time energy optimization.

Temperature and the amount of sunlight affect the power output of solar PV systems. MPPT is a control technique used to maintain optimal performance from the PV array. The P&O method is among the most widely used MPPT algorithms. Because it is straightforward, user-friendly, and compatible with real-time embedded systems like FPGAs, it is widely used.

Figure 6.13 shows a Simulink-based model of a solar PV system integrated with a DC-DC boost converter and a P&O MPPT algorithm. This simulation replicates a real-time solar charging setup for charging EVs, using environmental inputs such as irradiance and temperature. The model illustrates

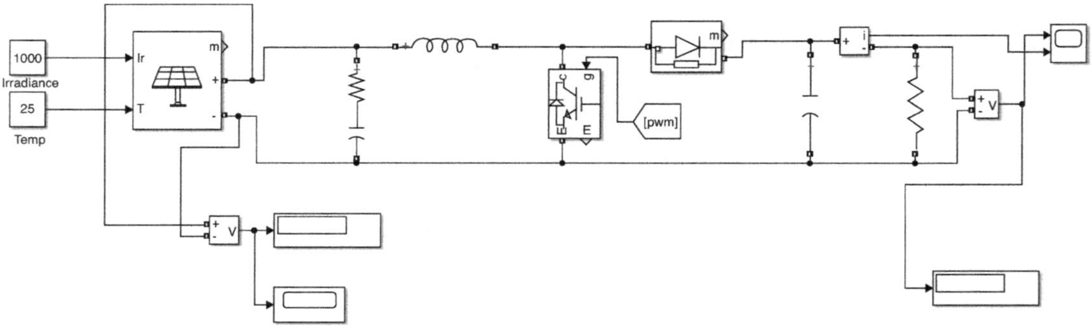

FIGURE 6.13 Diagram of a Simulink model of PV system with boost converter and FPGA-based P&O MPPT control.

how solar energy is collected, converted, and regulated to supply a controlled electrical load with PWM-driven switching and feedback. At the far left, two constant blocks mimic environmental conditions. Irradiance at 1000 W/m² represents peak solar radiation during full sunlight. Temperature at 25°C denotes a standard ambient temperature for PV testing.

These values feed into the PV array block, which models how a solar panel behaves electrically. The array generates DC voltage and current based on the environmental inputs and internal features like the number of cells and series resistance. Next, the circuit includes a filter made of a resistor and a capacitor in parallel. This filter cuts down high-frequency noise and sudden changes in the panel's DC output. It ensures that the voltage entering the DC-DC converter remains stable and dependable. The following stage has a boost converter that increases the voltage to the needed levels. This converter includes an inductor, a switching MOSFET, and a freewheeling diode. These key components work together to effectively raise the DC voltage gathered from the solar panels, making it suitable for later power conversion or grid connection.

6.3.3 Role of the Boost Converter in the System

In a PV system, the voltage from solar panels can change greatly due to factors like solar irradiance, temperature, and the specific setup of the PV array. Often, this raw DC voltage isn't at the right level for an inverter to convert it efficiently into AC power or for direct use on a high-voltage DC bus. This is where the boost converter is essential. It acts as a middle step to:

1. **Maximize Power Extraction**: By controlling the input impedance, the boost converter helps the PV array run at its MPP. This way, it extracts the most power possible under different environmental conditions. An integrated MPPT algorithm within the converter's control usually achieves this.
2. **Provide a Stable Higher Voltage**: It raises the fluctuating lower DC voltage from the solar panels to a stable, higher DC voltage level. This stable and higher voltage lets the downstream inverter work more efficiently and reliably, often making its design simpler.
3. **Optimize Grid Integration**: For on-grid systems, a higher DC link voltage can boost the efficiency of the grid-tie inverter and create a more compact, cost-effective overall system.

6.3.4 Detailed Working Principle of the Boost Converter

The boost converter works by storing energy in an inductor and then releasing it at a higher voltage. It uses rapid switching, usually controlled by a PWM signal from a dedicated controller. The operation can be divided into two distinct modes within each switching cycle:

Mode 1: Switch ON (Inductor Charging)

During this phase, the control signal turns the switching MOSFET ON (closed).

- **Current Path**: A low-resistance path opens from the input voltage source (the solar panel output) through the inductor and the conducting MOSFET to the ground.
- **Energy Storage**: The inductor, in series with the input, has a linearly increasing current as it stores energy in its magnetic field. The rate of current increase is directly related to the input voltage and inversely related to the inductor's inductance.
- **Diode Blocking**: At this time, the freewheeling diode is reverse-biased. The output capacitor, which maintains the higher output voltage, cannot discharge back into the input circuit because the diode blocks current flow in that direction. The load continues to draw current from the energy previously stored in the output capacitor.

Mode 2: Switch OFF (Inductor Discharging & Voltage Boost)

When the switching MOSFET is turned OFF (open), the current path through the switch is suddenly interrupted.

- **Inductor Voltage Spike**: The inductor, resisting the sudden change in current, quickly collapses its magnetic field. This creates a voltage across itself with a polarity that adds to the input voltage. This combined voltage (Vin+VL) appears across the inductor and diode.
- **Diode Conduction**: This much higher combined voltage forward-biases the freewheeling diode, allowing current to flow.
- **Energy Transfer to Output**: The stored energy from the inductor, along with the immediate input energy, is now sent through the diode to charge the output capacitor and power the connected load. The output capacitor smooths this pulsed current, providing a stable, higher DC output voltage.
- **Inductor Current Decrease**: As the inductor releases its stored energy, its current linearly decreases until the next ON cycle.

By carefully controlling the duty cycle (the ratio of the ON-time to the total switching period) of the MOSFET, the boost converter can manage the output voltage to the desired level. A longer ON-time allows more energy to be stored in the inductor, leading to a higher output voltage during the OFF-time, as long as the converter works in continuous conduction mode. This controlled and effective voltage increase is crucial for improving the performance and integration of solar energy systems.

The P&O MPPT algorithm finds the MPP by measuring the output voltage and current of the PV array and adjusting the PWM duty cycle accordingly.

6.3.5 Working Principle of the P&O Method

The P&O algorithm monitors how output power changes with slight adjustments in operating voltage or current. The main idea is to modify the PV operating voltage and observe the response of the output power. If the power increases, the algorithm continues adjusting in that direction. If the power decreases, the algorithm reverses its direction. This cycle repeats, helping the system find and maintain the MPP, where voltage and current yield the highest output.

Mathematically, if $P(k)$ is the power at time k and $P(k - 1)$ is the power at the previous iteration, then if $P(k)$ is greater than $P(k - 1)$ and $V(k)$ is greater than $V(k - 1)$, the voltage should be increased (keep direction). If $P(k)$ is less than $P(k - 1)$ and $V(k)$ is greater than $V(k - 1)$, then decrease the voltage (reverse direction). This simple logic ensures that the algorithm consistently moves toward greater power output until it reaches the MPP.

In this system, the P&O MPPT algorithm operates on an FPGA. FPGAs excel at parallel processing and fast computing, making them ideal for real-time energy management tasks. The FPGA executes the MPPT logic as an FSM, which performs these tasks in each cycle: acquisition, where real-time voltage and current from the PV panel are converted to digital form using ADCs and sent to the logic core.

The instantaneous power is calculated as $P(k)$ equals $V(k)$ multiplied by $I(k)$. The new power is then compared with the power from the last iteration, and the adjustment direction is updated. The PWM module in the FPGA modifies the DC-DC converter's duty cycle and adjusts the panel voltage accordingly. Power, voltage, and direction values are recorded for the next cycle. This fast feedback loop keeps the PV system operating at or close to its maximum efficiency.

In the model, voltage and current sensors supply real-time values to a P&O logic block, which adjusts the PWM signal sent to the gate terminal of the MOSFET. This control loop allows the boost converter to work at the MPP of the solar array despite changing conditions.

On the right side of the model from Figure 6.13, the output from the boost converter connects to a resistive load, simulating a battery or EV interface. The system constantly monitors the voltage and current across the load. A scope block shows the system's behavior, including voltage response, power output, and convergence to the MPP. These outputs are vital for evaluating the performance of the MPPT algorithm and the converter operation in the simulation.

Feedback and Control Monitoring: Various display and scope blocks are used throughout the model to check PV voltage and current and to visualize output power. Track the evolution of the control signal. This feedback is crucial for debugging and enhancing the algorithm's responsiveness and stability.

6.3.6 Things to Consider During Hardware Implementation

6.3.6.1 Selection criteria for PV modules in EV charging applications

When it comes to solar panels, one common question is which type of solar cell is better: Monocrystalline or polycrystalline?

Solar PVs convert sunlight directly into electric current at the junction of two materials. This occurs through the PV effect, where photons are absorbed and electrons are released. Solar energy consists of photons, which are tiny packets of electromagnetic energy, and the materials that exhibit the PV effect are known as solar cells or PV cells. These cells are made from semiconductor materials like silicon, which are also used in the Integrated Circuit (IC) design industry.

In solar cells, a thin semiconductor layer is doped to create an electric field, so that, when the sunlight hits the solar cell, it excites the electrons from the atoms in the semiconductor layer. This excitation of electrons leads to the flow of electric current. This electricity can power devices like lights or tools. The first PV module was created by Bell Laboratories in the year 1954. In modern days, such PV panels are categorized into two types based on the silicon structure used during manufacture. It is important to understand the differences between these two common crystalline silicon technologies for appropriate application (Table 6.5).

- **Monocrystalline Panels:** Due to the cylindrical nature of the pure silicon ingot, when slicing repolycrystallized solar panels, these rounded edges create small, visible gaps between the cells. The uniform crystal structure causes light to interact consistently, giving monocrystalline panels a uniform dark, often black, appearance. This distinctive look, combined with the rounded-edge squares and small inter-cell spaces, makes them easily recognizable.
- **Polycrystalline Panels**: Unlike monocrystalline wafers, polycrystalline wafers are cut from cubic ingots and therefore do not have rounded edges, resulting in full square cells.

TABLE 6.5 The difference between monocrystalline and polycrystalline solar panel

	MONOCRYSTALLINE SOLAR PANELS	*POLYCRYSTALLINE SOLAR PANELS*
Material	Single silicon crystal	Multiple silicon crystals fused together
Appearance	Black or dark blue	Blue in color with sharp edges
Conversion efficiency	15%–23%	13%–16%
Space efficiency	More efficient	Less efficient
Temperature coefficient	–0.3%/°C to –0.5%/°C	–0.3%/°C to –1%/°C
Lifespan	30+ years	25+ years
Recyclability	Yes	Yes
Cost	Higher cost	Lower cost

The presence of multiple silicon crystals within each cell causes light to interact differently, giving polycrystalline panels a characteristic bluish hue and a speckled appearance. Their identification is simplified by this bluish tint and the absence of rounded edges. Both types offer various back sheets and frame colors to suit aesthetic preferences.

6.3.6.2 Manufacturing process of the solar panels

The journey from raw silicon to a functional solar panel begins with the creation of highly pure silicon wafers.

6.3.6.2.1 Silicon purification: from quartzite to solar-grade silicon

The first step that is involved in the process of manufacturing a monocrystalline silicon cell is done by extracting pure silicon from the quartz sandstone or quartzite to make metallurgical silicon. This is done using special ovens that melt the silica and carbon at temperatures over 1400°C, leaving behind 98% to 99% pure silicon, also known as metallurgical silicon. Even though this metallurgical silicon has high purity, it is not pure enough to be used in PV panels, as even trace impurities can significantly impact efficiency. Thus, further purification is necessary. The next step is to purify the metallurgical silicon using the Siemens process.

This process uses the powdered metallurgical silicon and exposes it to a reactor filled with HCl at higher temperatures, leading to the formation of trichlorosilane gas. This gas is then cooled and liquefied for distillation. Distillation is the process of purifying a liquid by evaporating it and then condensing the vapors to separate the impurities. For example, by boiling seawater and condensing the vapor, pure water can be obtained, as the salt remains behind. Using the same method, the trichlorosilane liquid is evaporated and then cooled, removing the impurities with higher and lower boiling points, such as calcium and aluminum. After the distillation process, the purified trichlorosilane liquid is moved to another insulated reactor with a hot rod and is mixed with the hydrogen gas. This mixture is again vaporized at a temperature of about 1500°C. Due to the intense heat and the presence of hydrogen gas, the chlorine atoms dissolve, leaving behind silicon with 99.9999% purity. This highly pure silicon is essential for high-performance solar cells.

6.3.6.2.2 Monocrystalline silicon ingot production

What sets monocrystalline cells apart from polycrystalline cells is that the monocrystalline panels are made from a single silicon crystal. Creating this single or homogeneous, pure silicon was quite difficult until the development of the Czochralski process, named after the Polish scientist Jan Czochralski. The process begins by dipping a small rod of pure single-crystal silicon into molten silicon held in a crucible. After dipping the rod, it is slowly pulled and rotated upward simultaneously. This controlled movement helps reduce the effect of convection currents in the molten silicon, ensuring uniform crystal growth. As the crystal seed is pulled up, the liquid silicon gradually solidifies over approximately four days, forming a large, homogeneous cylindrical single crystal of silicon, also known as a silicon ingot. The final size and quality of the silicon ingot depend on three critical factors: the temperature gradient within the melt, the cooling rate, and the rotation speed of the seed crystal.

6.3.6.2.3 Slicing the ingot

Once the ingot is formed, the next crucial step is to slice it into very thin wafers. This is achieved using a specialized wire saw, which precisely cuts the ingot into slices typically around 1 mm (or 0.0393 inches) thick. After cutting, the wafers undergo polishing and washing to remove any dust, dirt, or scratches, ensuring a pristine surface for subsequent processing.

6.3.6.3 Improving the wafers for enhanced performance

The initial flat surface of the silicon wafer can cause a significant portion of incident light to reflect away, thereby reducing efficiency. To counteract this, manufacturers implement surface treatments:

- **Roughening and Etching:** The wafers' surfaces are intentionally roughened and etched. This textured surface allows light rays to refract multiple times within the silicon, increasing the probability of photon absorption and significantly improving the panel's efficiency by minimizing light reflection.

6.3.6.4 Creating the P-N junction by diffusion process

Silicon wafers, in their base form, are typically doped with elements like boron, making them positively charged (p-type material). For a solar cell to generate electricity, a p-n junction is essential.

- **Phosphorus Doping**: A negatively charged layer of phosphorus is added to each wafer. The wafers are then moved to specialized ovens, heated to approximately 1652°F (900°C), to facilitate the diffusion of phosphorus into the silicon lattice.
- **N-Type Layer Formation**: The introduction of phosphorus creates a powerful 1652°F p (negatively charged) layer on the surface of the p-type wafer. This forms the critical p-n junction, which is responsible for separating electron-hole pairs generated by sunlight, thereby enabling electricity conduction and significantly increasing the cell's efficiency.

6.3.6.5 For improved conductivity

To ensure efficient collection and transport of the generated electricity, measures are taken to enhance the wafer's conductivity:

- **Silver Alloy Application**: A highly conductive silver alloy is pressed onto the front surface of the wafer. This metallic grid acts as an electrical contact, minimizing resistance and ensuring that the generated power is efficiently transported away from the cell, further improving the overall conductivity of the monocrystalline cell.

6.3.6.6 Assembling monocrystalline solar panels

The final stage in manufacturing monocrystalline panels is assembly.

- **Cell Arrangement**: Each monocrystalline solar panel is composed of 32 to 96 individual pure crystal wafers. These cells are meticulously arranged in rows and columns and interconnected electrically.
- **Power Output Determination**: The total number of cells within a panel directly determines its overall power output. More cells generally equate to higher power generation.

6.3.6.7 Making of polycrystalline solar panels

Polycrystalline solar panels, also known as multi-crystalline or many-crystal solar panels, are also made from pure silicon but differ significantly in their initial crystal growth process.

- **Fragmented Silicon Crystallization**: Unlike monocrystalline production, where a single, homogeneous cylindrical crystal is slowly pulled (Czochralski process), for polycrystalline panels, the molten silicon is allowed to cool and solidify in a controlled manner within cubic-shaped growth crucibles. This process results in the formation of multiple silicon crystal fragments rather than a single continuous crystal.

- **Wafer Production**: After solidification, these multi-crystalline ingots are cut into thin wafers and then undergo similar subsequent steps as monocrystalline wafers: polishing, surface improvement (roughening/etching), diffusion (p-n junction creation), and assembly into panels.

6.3.6.8 Efficiency

Solar panel efficiency quantifies how effectively a cell converts incident sunlight into usable electricity. A higher efficiency means more power generated per unit of surface area.

- **Monocrystalline Panels**: These cells boast the highest efficiency among crystalline silicon types, typically ranging from 15% to 20% in commercial products, with ongoing research pushing these figures even higher. For instance, in 2019, the National Renewable Energy Laboratory achieved a record-setting 47.1% efficiency with a six-junction solar cell in laboratory conditions.
- **Polycrystalline Panels**: Due to the presence of multiple crystal boundaries within each cell, electron movement is slightly impeded, leading to marginally lower electricity generation efficiency. Most polycrystalline PV cells exhibit efficiencies between 13% and 16%. While lower than monocrystalline, this is still a very respectable range, and advancements continue to improve their performance.

6.3.6.9 Temperature coefficient

The temperature coefficient is a critical, an often-overlooked metric that indicates how well a solar cell performs as its temperature rises. Specifically, it quantifies the percentage of efficiency loss for every degree Celsius increase in temperature above standard test conditions (25°C).

- **Monocrystalline Panels**: Typically have a temperature coefficient ranging from approximately −0.3%/°C to −0.5%/°C. This means that for every 1°C (or 1.8°F) rise in temperature, a monocrystalline solar cell will temporarily lose 0.3% to 0.5% of its efficiency.
- **Polycrystalline Panels**: Generally exhibit a slightly higher (more negative) temperature coefficient than monocrystalline ones, often in the range of −0.3%/°C to −1%/°C. This implies that polycrystalline panels will experience a greater temporary efficiency loss when operating in higher temperatures, making them marginally less optimal for extremely hot climates compared to monocrystalline panels.

6.3.6.10 Expected lifespan

The lifespan of a solar cell is primarily indicated by its degradation rate, which is the yearly percentage loss in energy production. Most solar panels are designed with a degradation rate between 0.3% and 1%.

- **Monocrystalline Solar Panels**: Typically have a low annual efficiency loss, ranging from 0.3% to 0.8%. For a panel with a 0.5% degradation rate, it would operate at 95% efficiency after 10 years and 90% efficiency after 20 years. Most monocrystalline panels come with performance warranties of 25 or 30 years, but their operational lifespan can often extend to 40 years or more.
- **Polycrystalline Solar Panels**: Possess a slightly higher degradation rate than monocrystalline panels, leading to a marginally faster loss of efficiency over time. Nevertheless, they still offer a robust lifespan, typically ranging from 20 to 35 years, and often beyond.

6.3.6.11 *Recyclability*

Both monocrystalline and polycrystalline solar panels are largely recyclable, contributing to a more sustainable energy ecosystem.

- **Monocrystalline Solar Panels**: The main components are highly recyclable:
  - **Monocrystalline Cells**: Approximately 85% of the silicon wafers can be recycled.
  - **Glass**: Almost 95% of the glass can be reused.
  - **Metal Parts**: 100% of the metal components are recyclable.
- **Polycrystalline Solar Panels**: Similar to monocrystalline, around 90% of all materials used in the manufacturing of polycrystalline cells are recyclable. Projections suggest that by 2030, approximately 45 million new modules could be manufactured using recycled materials, representing an economic value of 380 million USD.

6.3.6.12 *Cost*

The initial cost is a significant factor in solar panel selection. Manufacturing and material differences between monocrystalline and polycrystalline solar panels significantly affect performance, temperature coefficients, efficiency, and cost. Monocrystalline panels offer higher efficiency but are more expensive, while polycrystalline panels are more cost-effective but slightly less efficient.

- **Monocrystalline Solar Panels**: Despite their numerous advantages, their primary disadvantage is the higher initial cost. The manufacturing process for monocrystalline cells is more energy-intensive and generates a larger amount of silicon waste, contributing to higher production expenses.
- **Polycrystalline Solar Panels**: Offer a lower cost per watt compared to monocrystalline panels, making them a more budget-friendly option. This is attributed to a less energy-intensive manufacturing process that produces less silicon waste. Interestingly, polycrystalline panels can sometimes be made from the leftover silicon fragments from monocrystalline production, further reducing waste. However, it is crucial to note that though cheaper, polycrystalline panels require more space to generate the same amount of energy, making them less space-efficient.

6.3.7 Sizing of Solar Arrays for EV Load Profiles

PV array sizing is the process of determining the optimal number and configuration of solar panels required to meet a specific energy demand. For EV charging stations, this process is particularly critical and complex due to the dynamic nature of EV charging loads and the intermittent nature of solar energy generation. Accurate PV array sizing ensures the system is both economically viable and technically capable of providing reliable and efficient charging services.

For example, on a clear and sunny day, a 1 kWp PV array received six peak sun hours (PSHs). Total loss (de-rating factor) in the system is estimated at 0.70 (70%).

Expected output can be determined as follows:

$$\begin{aligned}\text{Expected output} &= \text{Peak sun hours} \times \text{Peak power output} \times \text{total derating factor}\\ &= 1\,\text{kWp} \times 6\,\text{hours/day} \times 0.70\\ &= 4.2\,\text{kWh per day}\ (1\text{st year})\end{aligned}$$

Now considering degradation of module as per the indicative profile above (example only, actual degradation of module will be based on module quality and climatic conditions)

Energy generation:

= 3.83 kWh per day (on 10th year)
= 3.39 kWh per day (on 25th year)

$$W_{pv} = \left(E/\text{PSH} * \text{System efficiency}\right)$$

W_{pv} = peak wattage of the array
E = daily energy requirement
PSH: average daily number of peak sun hours in the design month for the inclination and orientation of the PV array
ηSys = total system efficiency

The month that the system is designed for is the month of the lowest average daily solar radiation during the operational period of the system.

The number of peak hours is for the inclination and orientation of the PV array. If the only information available is for solar radiation in a horizontal plane, then a tilt and orientation correction factor should be applied.

6.3.7.1 Figure 6.14: wiring solar panels in a series circuit

Connect the positive terminal of the first solar panel to the negative terminal of the next one as shown in Figure 6.14.

Example: If you had four solar panels in a series and each was rated at 12 V and 5 amps, the entire array would be 48 V at 5 A.

6.3.7.2 Wiring solar panels in a parallel circuit

Connect all the positive terminals of all the solar panels together and all the negative terminals of all the panels together, as shown in Figure 6.15.

Example: If you had four solar panels in parallel and each was rated at 12 V and 5 A, the entire array would be 12 V at 20 A.

FIGURE 6.14 A pictorial diagram representing the serial connection of solar PV panels.

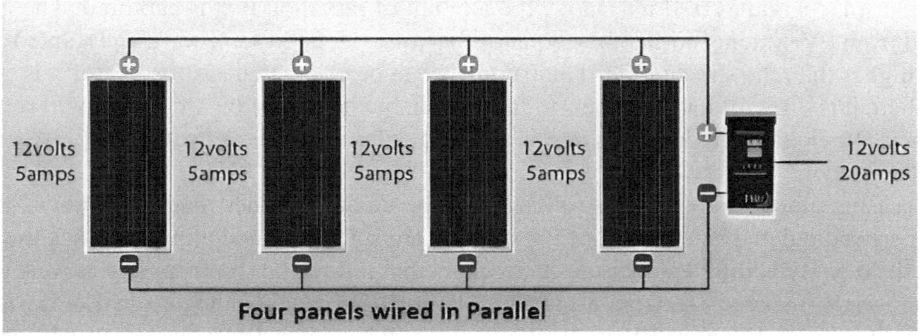

FIGURE 6.15 A pictorial diagram representing the parallel connection of solar PV panels.

6.3.8 Factors Affecting PV Module Performance

6.3.8.1 Location

The first step in developing a PV system is location. For any PV system to be financially viable, PV modules' access to sunlight is essential. Latitude is the primary factor.

6.3.8.2 Solar irradiance

It measures the amount of power or solar intensity, expressed in watts per square meter, that falls on a level surface or that reaches your position. A system's typical monthly and annual energy production can be predicted using historical, standardized weather data because weather patterns are often consistent over time.

Maps of solar resources that illustrate the amount of energy that reaches the panels' surface are available. Standardized maps that display the number of standard sunshine hours that can be calculated throughout a month or a year are used to portray the data.

6.3.8.3 Solar insolation

The amount of solar irradiance that reaches a PV surface at any given moment is known as solar insolation.

The amount of solar energy that is accessible at a specific place is measured in $kWh/m^2/day$. The term PSH is frequently used to describe this. For instance, a location's PSH will be five hours if its solar radiation is 5 $kWh/m^2/day$. Now, without accounting for losses, a 1 kW solar panel installed there would generate $1 \text{ kW} \times 5\text{h} = 5 \text{ kWh}$ of electricity daily. The module output will increase with stronger sunshine. Lower current output is the effect of less sunshine. Variations in the intensity of the sunlight do not significantly alter voltage.

6.3.8.4 PV Orientation and sun angle

The orientation of a solar panel is the direction it faces. The solar array's orientation is crucial because it influences how much sunlight strikes the array and, in turn, how much power it can generate. The tilt angle, or the angle between the solar panel's base and the horizontal, and the orientation typically comprise the direction the solar module is facing, or due south. Due to the sun's movement, the amount of sunlight that reaches the array also changes with the time of day.

Installing solar modules will maximize the amount of radiation that is captured. The best results are obtained from PV systems north of the equator that are orientated southward and slanted at an angle 15 degrees higher than the site latitude. The PV array can be orientated east or west if it is placed on a structure where it is difficult for the panels to face south, but never northward as this will result in very low efficiency. When a PV module's surface is perpendicular to the sun's beams, it operates at its peak power, or efficiency.

An increasing amount of energy is reflected by the modules rather than absorbed as the rays go away from perpendicular. The majority of PV systems are stationary and unable to track the sun's path throughout the day. By adding PV modules to trackers that follow the sun from east to west throughout the day (single-axis trackers) and from north to south during seasonal changes (dual-axis trackers), the output can be increased. For most PV applications, this is not standard practice because it can be costly. To capture most of the solar energy throughout the year, PVs should be angled toward the sun's average elevation, which is equal to the latitude of the array's position. As an illustration, a system to generate the largest quantity of electricity over a year, a tilt angle of 15° to 35° can be used at a latitude of 25° throughout the year.

6.4 EV CHARGING

EVs are revolutionizing the transportation landscape, offering a cleaner, more efficient alternative to traditional gasoline-powered vehicles. EVs use electricity stored in rechargeable batteries to power electric motors for vehicle propulsion, significantly reducing or eliminating tailpipe emissions, which contribute to improved air quality and a reduction in greenhouse gas emissions [2]. They come in various types, including battery electric vehicles (BEVs) that run solely on electricity, plug-in hybrid electric vehicles (PHEVs) that combine a battery-powered electric motor with an internal combustion engine, and hybrid electric vehicles (HEVs) that use both an electric motor and a conventional engine but are not typically plugged in to charge [2].

6.4.1 Introduction to EV and Charging

EVs provide several advantages over conventional vehicles. One of the most notable benefits is their higher efficiency; electric motors convert a greater proportion of energy into vehicle movement, with efficiencies around 60% to 77%, compared to gasoline engines that typically convert only 17% to 40% of fuel energy into motion [2]. This efficiency translates into lower operating costs, with charging an EV generally costing less than fueling a gasoline vehicle. Additionally, EVs offer quieter operation and smoother acceleration, enhancing the driving experience. The increasing adoption of EVs is driven not only by technological advancements but also by growing environmental concerns and supportive government policies. The transportation sector is a significant contributor to global carbon emissions, accounting for over 15% of energy-related emissions worldwide, making the transition to EVs a critical element of decarbonization strategies [2]. Advances in battery technology, including improvements in energy density, lifespan, and safety, alongside cost reductions, are enabling more affordable and longer-range EVs, addressing earlier concerns about range limitations and vehicle cost. Charging infrastructure is a fundamental aspect of the EV ecosystem. EVs require access to reliable, accessible charging stations to meet their energy needs, whether at home, work, or public locations. There are three main levels of charging:

6.4.1.1 Charging

Level 1 charging uses a standard 120 V household outlet and delivers about 3–5 miles of range per hour of charging. It is typically slow and used in emergency or low-demand situations. Level 2 charging operates on 208–240 V power, similar to a clothes dryer outlet, providing 12–80 miles of range per hour of charging. Most home, workplace, and public chargers are Level 2, enabling full charges in a few hours [4]. DC Fast Charging (Level 3) uses DC to deliver rapid charging, typically adding 100 to 200+ miles of range in 30 minutes or less. These chargers are ideal for long-distance travel and are increasingly deployed along highways and high-traffic areas (Figure 6.16).

CHAdeMO standards promote broader interoperability [4]. Emerging technologies such as wireless (inductive) charging offer the convenience of charging without cables by transferring power through electromagnetic fields. Currently, wireless charging typically delivers power at rates comparable to Level 2 chargers and is more common for fleet and transit applications. Research continues into higher-powered wireless charging and in-motion charging solutions to further enhance convenience and reduce barriers to EV adoption [4].

The concept of integrating renewable energy sources like PVs (solar power) into EV charging infrastructure is gaining traction. Solar-powered charging stations reduce the carbon footprint of EV charging by supplying electricity from clean, sustainable sources. Solutions such as PV-based off-grid charging stations, controlled by sophisticated converters, and PV-powered V2G enabled charging stations are being developed, enabling EVs to not only consume renewable energy but potentially feed electricity back into the grid during peak demand. Smart charging solutions that optimize charging times, power delivery, and grid impact play an increasingly important role in managing energy demand and enhancing the user experience. Smart chargers can schedule charging during low-demand periods, communicate with utility grids, and integrate with EMSs to minimize costs and grid stress. This is essential as EV adoption scales, ensuring sustainable and efficient use of electricity resources [4].

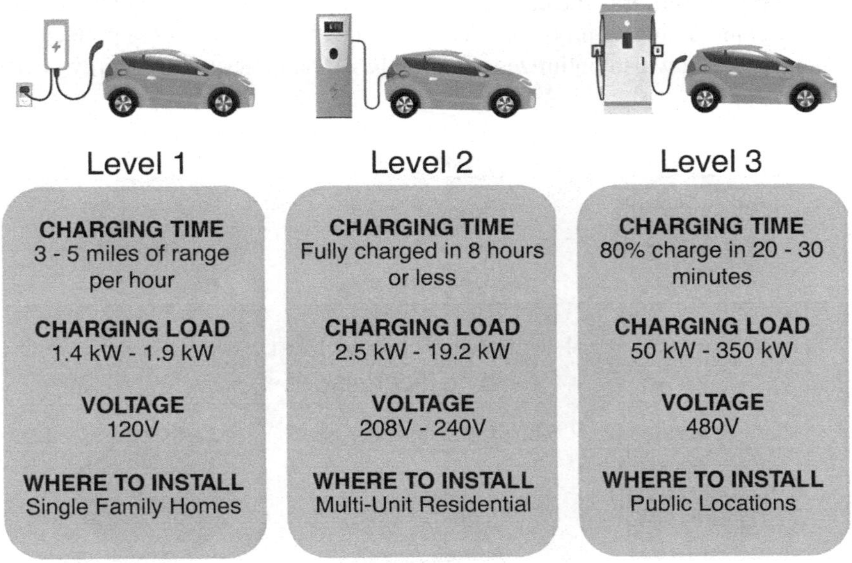

FIGURE 6.16 Charging connectors vary by region and manufacturer but commonly include the SAE J1772 for Level 1 and Level 2, and Combined Charging System (CCS), CHAdeMO, and Tesla's NACS for DC Fast Charging.

EVs are central to the global strategy of transitioning to sustainable transportation systems. They significantly reduce dependence on fossil fuels, cut greenhouse gas emissions, and offer long-term economic benefits through lower fuel and maintenance costs. As nations intensify efforts to combat climate change, EVs play a critical role in achieving net-zero carbon targets. The rapid evolution of battery technologies, particularly in lithium-ion and solid-state chemistry, is improving energy density, charging speed, and battery longevity. Simultaneously, the expansion of charging infrastructure from residential AC chargers to ultra-fast DC public charging stations is addressing range anxiety and improving accessibility [17]. Integration with renewable energy sources, such as solar and wind, is enabling cleaner charging and promoting grid resilience through V2G capabilities. Advancements in smart charging systems are enhancing load management by enabling DR and time-of-use optimization. Wireless and dynamic charging technologies are set to revolutionize convenience, allowing EVs to charge while parked or even in motion. Government incentives, stricter emission regulations, and increased automaker investment are accelerating EV adoption globally. Despite persistent challenges like high initial purchase costs and limited charging availability in some regions, continued technological innovation, supportive public policy, and consumer awareness are steadily overcoming these barriers. Collectively, these developments are driving a transformative shift toward a cleaner, more connected, and sustainable transportation future dominated by electric mobility [17] (Figure 6.17).

Figure 6.17 presents six methods: Inductive Charging, Ultra-Fast Charging, DC Fast Charging, Tesla-Superchargers, Bidirectional Charging (V2G), and Battery Swapping. These techniques are aimed at reducing the time required to recharge an EV's battery.

EV charging systems are composed of various interconnected components and technologies that supply electrical energy to recharge EV batteries in an efficient, safe, and reliable manner. The fundamental structure of an EV charging system includes several key components: a power supply unit, charging connectors and cables, control and communication units, and, in advanced setups, an EMS. The power supply unit is responsible for converting and conditioning electricity, typically rectifying utility-supplied AC power to regulated DC power when needed. Charging connectors and cables serve as the physical interface between the charger and the vehicle, enabling secure transmission of power and data. The control unit governs the charging process, handling voltage regulation, current control, implementation of communication protocols, and ensuring all necessary safety protections. Some systems also include an EMS, which optimizes charging based on battery SoC, energy tariffs, grid load,

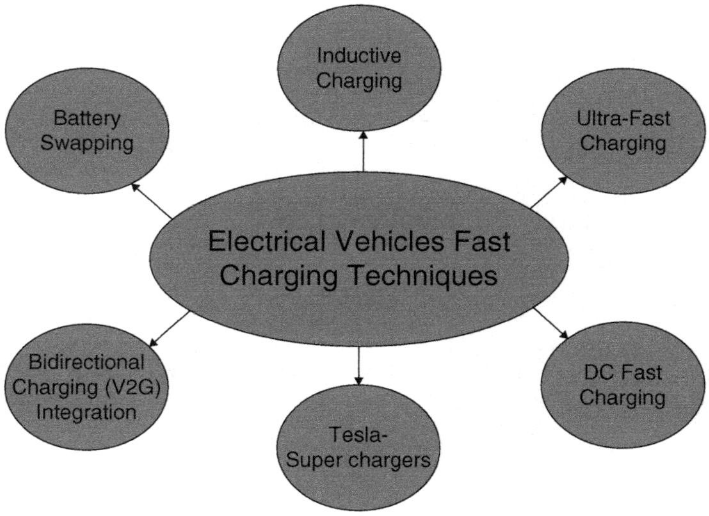

FIGURE 6.17 Various fast-charging techniques for electric vehicles (EVs).

or integration with renewable sources and energy storage systems. EV charging systems can be architecturally categorized based on where the power conversion from AC to DC takes place. In onboard chargers, the AC-to-DC conversion occurs inside the vehicle itself; the external charger simply supplies AC power. Conversely, off-board chargers perform this conversion externally, delivering DC power directly to the battery—this configuration is typical of fast-charging stations. Modern architecture increasingly incorporates smart charging features, support for renewable energy integration (e.g., solar PV), BESS, and bidirectional power flow capabilities that enable V2G operations. Such advanced systems not only charge EVs but also contribute to grid stability and energy optimization [17].

6.4.2 Major EV Charging Methods

EV charging technologies are broadly categorized by the type of current used and the infrastructure configuration, resulting in different levels of charging performance as per Figure 6.18 [17]:

1. Trickle Charging (Level 1 Charging) utilizes a standard 120 V AC household outlet and is the simplest and slowest form of EV charging. It typically delivers around 1.4–1.9 kW, adding about 4–5 miles of range per hour. Though ideal for overnight charging or vehicles with small battery capacities, it is not practical for vehicles with high daily driving needs. It requires no dedicated infrastructure and uses the standard charging cable provided with most EVs.
2. AC Charging (Level 2 Charging) employs a 208–240 V AC supply and significantly improves charging speed, delivering between 3.3 and 22 kW depending on the charger capacity and vehicle compatibility. It can add between 12 and 80 miles of range per hour. This type is widely used in residential, commercial, and public charging installations. The AC power is converted to DC within the vehicle via the onboard charger.
3. DC Fast Charging (Level 3 Charging) bypasses the onboard charger and delivers high-voltage DC (typically between 400 and 1,000 V) directly to the battery. These systems operate at power levels ranging from 50 kW to over 350 kW and can recharge an EV battery to 80% in approximately 20–30 minutes. DC fast chargers are ideal for public charging stations, especially in highway corridors and urban high-traffic zones. However, frequent fast charging can stress battery chemistry, potentially reducing long-term battery life.

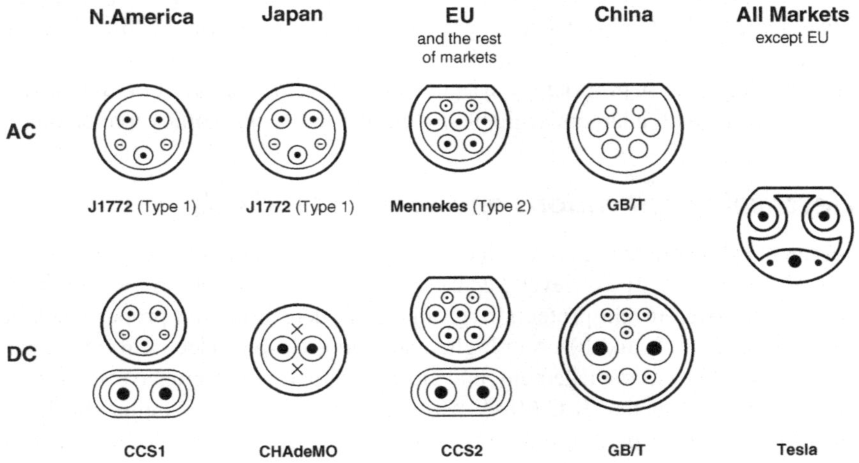

FIGURE 6.18 Types of connectors for EV charging.

6.4.2.1 Emerging and advanced charging technologies

Beyond the traditional methods, several innovative EV charging technologies are being explored and adopted [5]:

- Wireless charging uses inductive power transfer through magnetic fields between a ground-mounted charging pad and a receiver coil in the vehicle. Static wireless systems offer capacities up to 22 kW, while dynamic wireless charging—currently under experimentation—aims to charge vehicles in motion using embedded coils along roadways.
- Battery swapping enables the quick replacement of a depleted battery with a fully charged unit at dedicated stations. This significantly reduces downtime and is especially useful in fleet operations. However, widespread adoption is limited by the need for standardized battery form factors and significant infrastructure investment [18].
- V2G technology [4] allows EVs to not only receive power from the grid but also supply it back when needed. This two-way energy exchange helps in grid load balancing and enhances renewable energy utilization. V2G requires smart chargers, bidirectional inverters, and communication interfaces to operate effectively.

6.4.2.2 Connector types and compatibility

EV connectors are standardized across regions to ensure interoperability and safety. For AC charging, the most common connector types are:

- **Type 1 (SAE J1772)**: Used predominantly in North America and Japan, it supports single-phase AC charging and has five pins.
- **Type 2 (IEC 62196)**: Common in Europe and parts of Asia, this seven-pin connector supports both single-phase and three-phase AC charging.

For DC Fast Charging, several standards exist:

- **CHAdeMO**: A Japanese standard used primarily by Nissan, Mitsubishi, and other Asian manufacturers, supporting fast DC charging with separate communication and power pins.
- **CCS (Combined Charging System)**: A global standard integrating AC (Type 1 or Type 2) with two additional DC pins. CCS1 is prevalent in North America, while CCS2 is used in Europe and other regions.
- **Tesla Supercharger**: A proprietary system for Tesla vehicles, based on a modified Type 2 design in Europe, with adapters available for compatibility with other connectors.

6.4.2.3 Electrical specifications and safety standards

Charging equipment must adhere to defined electrical and safety standards. Level 1 chargers operate at 120 V, typically delivering 12–16 A. Level 2 chargers use 208–240 V and provide 16–80 A, requiring dedicated circuits with overcurrent protection, ground-fault detection, and proper insulation. DC fast chargers often operate at 480 V and 100 A or higher, and require robust electrical infrastructure including heavy-duty wiring, thermal management, and electromagnetic interference mitigation.

International standards such as IEC 61851, IEC 62196, and SAE J1772 define mechanical, electrical, communication, and safety parameters for EV charging systems. These standards ensure interoperability across vehicles and chargers, enforce safety protocols (like auto-disconnects and fault detection), and standardize communication interfaces for smart charging and grid interaction.

6.4.3 Charging Technologies

Figure 6.19 presents a classification of charging technologies for BEVs. Wireless charging is categorized into "Near-Field," "Medium-Field," and "Far-Field," representing current and future technologies and includes methods such as Inductive, Capacitive, Laser, and Microwave charging.

6.4.3.1 Wired charging

Wired charging [19] is the traditional and most widely deployed method for charging BEVs. It involves a direct physical connection between the electric grid or charger and the vehicle. This method is preferred due to its simplicity, mature infrastructure, and relatively higher efficiency compared to wireless alternatives. Wired charging is further subdivided into AC charging and DC charging, depending on the type of electrical current delivered to the vehicle.

6.4.3.2 AC charging

AC charging is commonly used for home and workplace applications where longer charging durations are acceptable. In AC charging systems, the power from the grid is supplied as AC, and the vehicle's onboard charger is responsible for converting this into DC that can be stored in the battery.

- **Single-Phase (1ϕ) Onboard Slow Charging**: This is typically found in residential settings. The system operates in the range of 3.3–7.2 kW, offering a slow but safe charging option. It is convenient for overnight charging and does not require special infrastructure beyond a standard outlet with an appropriate power rating. Vehicles can be charged over several hours, usually overnight, making it ideal for daily commuting needs [19].
- **Three-Phase (3ϕ) Onboard Fast Charging**: This is more common in commercial or industrial locations. With power ratings of up to 22 kW or more, this method offers significantly faster charging times compared to single-phase systems. It requires a three-phase electrical supply and is useful in environments where quick turnaround or shared charging stations are needed [19].

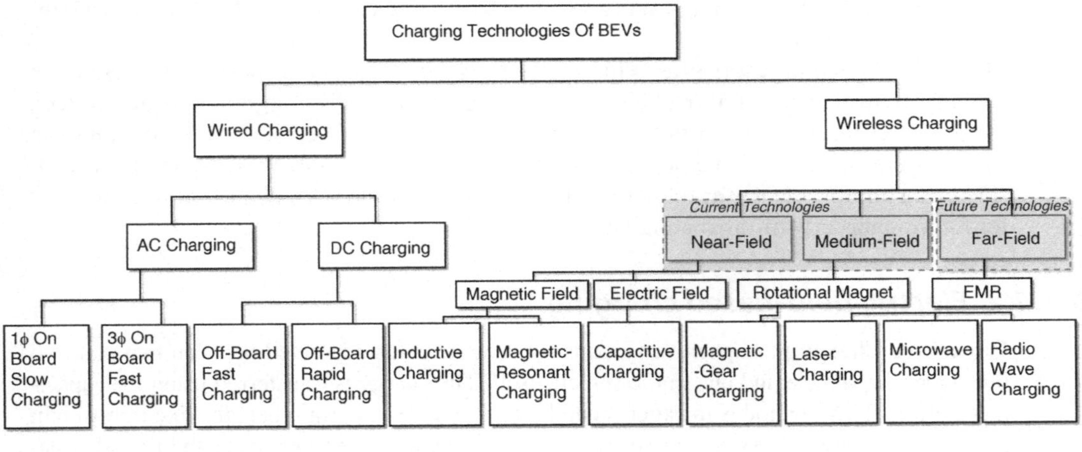

FIGURE 6.19 Represents the charging technologies.

6.4.3.3 DC charging

Unlike AC charging, DC charging delivers high-power DC directly to the battery, bypassing the onboard charger. This enables much faster charging, making it suitable for long-distance travelers, fleet operators, and high-usage scenarios.

- **Off-board Fast Charging**: These chargers are located outside the vehicle and handle the conversion from AC to DC externally [19]. Power levels are higher than typical AC chargers, providing faster recharge times. These are commonly installed at shopping centers, workplaces, and public parking facilities.
- **Off-board Rapid Charging**: This is among the fastest commercially available charging methods. With power levels that can exceed 100 kW and even go up to 350 kW, it allows charging up to 80% of the battery in less than 30 minutes. Such systems are ideal for highway service stations and are critical for enabling long-distance EV travel without excessive downtime [19].

6.4.4 Wireless Charging

Wireless charging, also known as inductive or contactless charging, offers the advantage of convenience and automation by eliminating the need for plug-in cables. This technology enables vehicles to charge by simply parking over a designated charging pad or even while in motion on specialized roads. It relies on electromagnetic fields to transfer power between the transmitter (usually on the ground) and the receiver (mounted on the vehicle).

Wireless charging technologies can be categorized based on the type of electromagnetic mechanism used to transfer energy. These include Magnetic Field, Electric Field, Rotational Magnet, and Electromagnetic Radiation (EMR)–based systems [19].

6.4.4.1 Magnetic field-based charging

- **Inductive Charging**: Inductive charging uses two coils—one in the charger and one in the vehicle—that create a magnetic field to transfer energy wirelessly. This method is already commercially deployed in some luxury EVs and public transit buses. It is a near-field technology, meaning it works only when the coils are closely aligned (typically within a few centimeters) [19]. Although it is safe and convenient, efficiency can suffer due to misalignment or vertical gaps.
- **Magnetic Resonant Charging**: This method also uses magnetic fields but operates at resonant frequencies that allow better energy transfer even when alignment is not perfect. Magnetic resonant systems offer more flexibility in terms of vehicle positioning and can support both static and dynamic charging (charging while the vehicle is moving). It is considered an improvement over traditional inductive systems and is gradually gaining attention for future smart city infrastructure [19].

6.4.4.2 Electric field-based charging

- **Capacitive Charging**: Capacitive systems use electric fields generated between two plates—one under the vehicle and one in the vehicle itself. The energy is transferred using the capacitance effect. This method is in experimental stages and offers certain benefits like lightweight and easy integration. However, challenges include limited power transfer capability and sensitivity to misalignment, making it less practical for current BEV energy demands.

6.4.4.3 Rotational magnet-based charging

- **Magnetic Gear Charging**: This innovative concept involves the use of rotating magnetic fields to generate electricity and charge the vehicle. The idea mimics mechanical gear systems but is implemented magnetically. While this approach holds theoretical promise, it is still in the research and prototype stage due to complexities in implementation and low efficiency under current designs.

6.4.4.4 EMR-based charging

These systems use beams of energy in the form of EM waves to transfer power over longer distances without physical contact. While offering far-field charging capabilities, they are still under heavy research due to safety, cost, and efficiency concerns.

- **Laser Charging**: Laser systems can focus energy beams over long distances to charge vehicles. The receiver converts the laser into usable electrical energy. Although promising in terms of range and speed, laser systems require line-of-sight, and safety concerns due to the intensity of the laser beam are significant.
- **Microwave Charging**: This uses microwave radiation to transfer energy wirelessly over medium distances. It is theoretically suited for dynamic charging applications where vehicles are in motion. However, microwave systems suffer from low conversion efficiency, regulatory challenges, and potential health risks due to radiation exposure.
- **Radio Wave Charging**: Operating at even lower frequencies than microwaves, radio wave systems can transfer small amounts of energy over very long distances. Though not currently suitable for full-scale EV charging, they may find applications in charging low-power sensors or auxiliary systems. Their future integration into automotive systems depends on technological breakthroughs in receiver efficiency and power conversion [19].

6.4.4.5 Wireless charging field range classifications

Wireless systems are also categorized by the operational distance between the charger and the vehicle, which significantly affects feasibility, safety, and efficiency.

- **Near-Field**: These systems (e.g., inductive and magnetic resonant charging) operate over very short distances, typically less than a few centimeters. They are already deployed commercially and offer a good balance of efficiency and safety. Current EVs with wireless capabilities use this class of technology.
- **Medium-Field**: Medium-range systems offer charging over distances of several meters. These are under development, especially in the form of improved magnetic resonant or microwave systems [19]. They are promising for fleet vehicles and autonomous parking systems.
- **Far-Field**: Far-field systems include laser and radio wave charging, capable of transmitting energy over long distances (tens of meters or more). While these are highly futuristic, they may one day allow BEVs to charge while in motion or in remote areas without physical infrastructure [19]. However, significant technological, safety, and regulatory hurdles remain (Table 6.6).

6.4.5 Setting up an EV Charging Station

The EV charging station system as mentioned in the Figure 6.20 can be conceptually divided into three main components [20] (Figure 6.20):

TABLE 6.6 Comparison of various electric vehicle (EV) charging methods based on their speed, convenience, and current level of adoption or technological maturity

METHOD	*SPEED*	*CONVENIENCE*	*CURRENT STATUS*
1ϕ AC slow charging	Slow	High	Widely used in homes
3ϕ AC fast charging	Medium	Medium	Used in public places
DC rapid charging	Very fast	Low (due to cost and heat)	Growing rapidly
Inductive charging	Medium	High	Current wireless tech
Resonant charging	Medium	High	Increasing adoption
Capacitive charging	Low-medium	Medium	Experimental
Laser charging	High	High	Research stage
Microwave/radio wave	Low	Very high	Long-term future tech

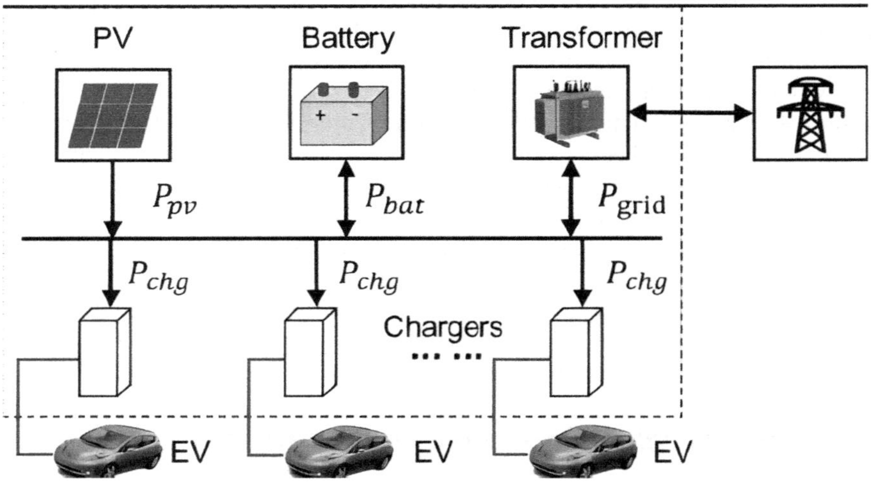

FIGURE 6.20 Represents configuration of EV charging station.

- Power generation/storage sources
  - **PV Panels**: Convert solar energy into electrical energy Ppv and supply DC power.
  - **Battery Storage**: Stores energy Pbat for use during peak demand or when PV generation is low.
- Power conversion and chargers
  - Each source feeds power to chargers through power electronics to regulate and control the charging power Pchg delivered to EVs.
  - Chargers convert electrical power as necessary (AC to DC or DC to DC) for battery recharging in EVs.
- Grid interface
  - A transformer connects the charging system to the main utility grid, allowing flow of power Pgrid in both directions.
  - The grid acts as a backup power source or a sink for excess energy (e.g., from PV or battery).

This architecture supports an integrated, hybrid energy setup, leveraging renewable generation, energy storage, and grid power.

6.4.5.1 *Site selection and preliminary planning*

The initial phase of setting up an EV charging station involves strategic site selection and thorough preliminary planning. It is crucial to choose a site that is easily accessible, offers adequate parking space, and has proximity to existing grid connectivity for seamless electrical integration. A detailed site survey should be conducted to evaluate the condition and capacity of the existing electrical infrastructure, including the availability of utility power. Environmental considerations such as local constraints, solar irradiation levels (particularly if PV systems are to be integrated), and the future scalability of the station must be assessed. Furthermore, it is essential to ensure that the proposed installation complies with local building codes, electrical safety standards, and zoning regulations, which will facilitate smoother project approval and reduce future legal or operational issues [21].

6.4.5.2 *Electrical system design and load analysis*

In the design phase, developing a clear power flow diagram is vital. The typical schematic illustrates three major power sources: Ppv from solar panels, Pbat from battery storage, and Pgrid from the utility grid. These sources converge to supply energy to EV chargers, which ultimately deliver power Pchg to the EVs. A detailed load estimation must follow, evaluating the total expected charging demand based on the number of chargers and predicted peak usage. This includes assessing simultaneous charging scenarios to determine worst-case load conditions. Power electronic converters play a crucial role in controlling energy flow from multiple sources while maintaining grid stability and ensuring compatibility with EV chargers. These converters should also allow modular expansion of the station in the future [21]. Additionally, MPPT algorithms are integrated to optimize the performance of PV systems by dynamically adjusting to changing solar input conditions. For battery energy storage systems, BMSs must be employed to ensure safe and efficient charge and discharge operations, monitor battery health, and prevent overcharging or deep discharging [21].

6.4.5.3 *Equipment selection*

The selection of high-quality and compatible equipment is fundamental to the success of a charging station. For PV systems, PV modules should be chosen based on the solar potential of the location, ensuring maximum energy capture [22]. If the grid or chargers require AC, suitable inverters must be included to convert the DC output of the solar panels into usable AC. Battery storage systems must be carefully selected, usually opting for lithium-ion batteries due to their high energy density and reliability. The battery capacity should be aligned with the site's load demand and desired energy autonomy. To manage energy efficiently between the batteries and chargers or grid, DC/DC converters with integrated charge controllers should be used [22]. Transformers and grid interfaces must be rated to handle the maximum expected power output with adequate safety margins. These transformers should be supported by protective devices such as circuit breakers, relays, and meters for safe grid synchronization and performance monitoring. In selecting EV chargers, the equipment must be compatible with the types of EVs expected to use the station and should support the appropriate charging levels—Level 2 AC chargers for slower applications and DC fast chargers for rapid charging. These chargers must also support standard communication protocols like CHAdeMO, CCS, or ISO 15118 to ensure interoperability and enable smart charging functions.

6.4.5.4 *Installation process*

The installation process begins with site preparation, which includes mounting structures for PV panels and laying concrete pads or enclosures for battery banks and transformers. The EV chargers are either pedestal-mounted or fixed to walls depending on the site layout and user access. Once physical installation is complete, detailed electrical wiring must be performed. This involves connecting PV arrays to inverters and routing the generated power to the battery system or directly to the chargers. Bidirectional DC/DC converters are used to link the battery system with the chargers, enabling both charging and discharging operations. The transformer's secondary output is wired to distribution panels that feed the chargers [22]. All cabling must be routed through conduits or protected channels, and cable management systems should be implemented to maintain safety and orderliness. Earthing and grounding must adhere strictly to national electrical codes (NECs) to prevent potential hazards. Overcurrent protection devices such as circuit breakers and Residual Current Device (RCD) or Ground Fault Circuit Interrupter (GFCIs) must be installed at critical points. All wiring and protective systems must comply with standards like the NEC, IEC standards, or other applicable local regulations.

6.4.5.5 System integration and communication

For seamless operation and scalability, the charging infrastructure must be integrated into a centralized control and monitoring system. The chargers should be networked and linked to a local control system or a cloud-based platform [22]. This facilitates real-time monitoring, load balancing across chargers, fault detection, and user authentication. Data communication is essential and should follow open protocols like the OCPP, enabling station management, remote diagnostics, and billing processes. If V2G functionality is supported, the communication system must allow bidirectional data flow between the charger and the vehicle. An EMS should be integrated to coordinate energy dispatch from PV panels, battery storage, and the utility grid based on current demand, weather conditions, and real-time electricity prices. Predictive algorithms within the EMS can schedule charging sessions efficiently, reduce peak demand charges, and improve overall system performance.

6.4.5.6 Testing and commissioning

Before the station is opened for use, thorough testing and commissioning must be performed. All electrical connections must be verified, and insulation resistance tests should confirm there are no leakage paths or faulty wiring. System control functions such as MPPT for the PV system, charge and discharge cycles of the battery storage, and proper synchronization with the grid must be tested for functionality and safety. Protection relays should respond accurately to fault conditions, and system recovery protocols must be validated. Chargers should be tested using actual EVs or load simulators to verify charging speed, current levels, and user interface functions. Additionally, communication links should be confirmed to be functional, ensuring that remote monitoring, user access, and backend data collection are fully operational. Safety systems, including emergency stops, fault indicators, and grounding checks, must be validated to comply with regulatory and operational standards.

6.4.5.7 Operation and maintenance

A well-maintained charging station is essential for longevity and reliable operation. The entire system should be monitored in real time via EMS dashboards, which provide insights into power usage, charging patterns, equipment status, and energy source contribution. Scheduled maintenance activities should include physical inspection of cables, electrical connectors, PV panel cleanliness, inverter and charger performance, and battery health checks. Software and firmware updates must be applied regularly to chargers, control systems, and EMS platforms to ensure compatibility with new EV models and grid requirements. Efficient customer support must also be in place to handle user authentication issues, billing queries, and general troubleshooting, ensuring a smooth and satisfactory user experience.

6.5 CONCLUSION

The solar-powered EV charging station achieves robust performance by intelligently coordinating PV generation with battery storage and the EV charging load. In bright sunlight the PV array is connected through an FPGA-based MPPT controller to capture maximal energy. The FPGA constantly adjusts the DC-to-DC converter to keep the panels at their optimal operating point, routing excess power directly to the EV or into the battery storage as needed. When sunlight is partial or fluctuating the FPGA dynamically balances power flows from both the PV array and the battery to meet charging demand. In darkness or low-generation periods the battery or grid supply as a backup provides energy to the EV charger, with the FPGA smoothly switching modes in real time. These transitions are seamless as the system maintains steady voltage and current to the EV outlet under all conditions. The FPGA-based control system significantly enhances the performance of solar-powered EV charging stations. By enabling near-optimal extraction of solar energy even amid fluctuations in irradiance and temperature the system boosts the overall energy conversion efficiency. The FPGA generates high-frequency highly precise PWM signals that drive advanced SiC or GaN power transistors with clean switching behavior. Performance metrics clearly show the advantages of using FPGA-based control. This approach allows for almost optimal extraction of solar energy even when sunlight levels and temperature fluctuate, which in turn improves the overall efficiency of the system. The FPGA produces high-frequency and precise PWM signals that ensure smooth and clean switching in the power electronics, especially when paired with advanced SiC or GaN transistors. This results in minimal energy losses during conversion. Meanwhile, the BMS keeps the SoC within safe limits by avoiding deep discharges or overcharging, which helps extend battery life. In everyday use these capabilities lead to efficient use of the available solar power with steady delivery of power to the vehicle and a reliable charging service that maximizes uptime and throughput.

The FPGA acts as the brain of the system, taking advantage of its hardware parallelism and flexibility to carry out complex control tasks with very low delay. At the heart of this is an MPPT module built directly into the FPGA logic such as a P&O algorithm implemented as an FSM. The FPGA continuously measures the voltage and current from the solar panels, calculates the power output, and instantly adjusts the converter's duty cycle to keep the panels operating at their MPP. Because these calculations and adjustments happen inside dedicated hardware, there is virtually no lag in the control loop, which allows the solar array to consistently run at peak efficiency. Besides MPPT the FPGA also generates the precise PWM signals that control the DC-to-DC converters and inverters responsible for managing the voltage between the solar panels, battery, and EV charger. The timing accuracy of these FPGA-controlled PWM signals is crucial as it ensures that energy transfer is efficient when waveform quality is high and switching losses are kept to a minimum. The FPGA also handles real-time monitoring and safety features. It receives digital data from sensors measuring voltage, current, temperature, and solar irradiance. The control logic constantly checks these measurements against safety limits. If it detects conditions like overcurrent, overvoltage, overheating, or other faults the FPGA can instantly shut down or isolate the affected components to protect the system. This built-in hardware-level fault detection helps prevent damage to the solar panels, batteries, and chargers, ensuring the system remains dependable over the long term.

Modern FPGA designs often include embedded CPU cores (or external microcontrollers) to handle communication and higher-level coordination. For example, the FPGA can host a soft-core or hard-core ARM processor that runs firmware to manage network interfaces and execute scheduling logic. This processor can implement EVSE communication protocols such as CAN bus for ISO 15118 or UART for simpler EV charging standards to coordinate charging sessions and enable bidirectional (V2G) signaling. It can also connect via Ethernet, SPI/I2C, or other fieldbuses to a building or grid EMS. In this hybrid architecture, the FPGA fabric offloads all critical timing-sensitive loops, while the embedded software handles supervision, data logging, and integration tasks. The result is a highly flexible control platform as one hardware setup can be reprogrammed as needed to optimize charging schedules, respond to grid signals, or update energy management policies.

The system's modular design makes it easy to scale. The same FPGA-based controller can be used for small setups or expanded to manage multiple solar panels, batteries, and EV chargers. For example, each parking stall could have its own simple FPGA controller or a more powerful FPGA with many I/O ports, and processor cores could handle several chargers at once. This flexibility means the system can grow from a home installation to a large commercial station without major redesign. Lower-end FPGA chips work well for basic stations, while advanced SoC FPGAs can support more complex features like machine learning for smarter management. Integration with the smart grid is another major strength. The station supports two-way power flow, allowing excess solar energy to be sent back to the grid and letting EV batteries return stored power during peak demand (V2G). The FPGA handles all necessary communication protocols and connects seamlessly to building or utility energy systems via standard networks like Ethernet, SPI, or UART. This enables participation in DR programs and real-time price-based charging adjustments, which helps balance grid load while optimizing costs and renewable energy use. While FPGAs cost more upfront than simple microcontrollers, their flexibility pays off over time. Firmware updates can add new features, improve safety, or support emerging charging standards, with no hardware changes needed. A single FPGA board can perform multiple functions by switching logic configurations, reducing hardware complexity. This modular upgradable design extends the station's life and makes it a smart investment. Research is ongoing to make these systems even better. Advanced control algorithms like adaptive MPPT or AI-driven energy management could help handle challenges like shading, sudden weather changes, or multiple EVs charging simultaneously. By using forecast data and grid signals, charging and storage can be scheduled proactively to maximize efficiency. New power electronics from improved converters to faster switches can also be leveraged thanks to the FPGA's precise control. Real-world pilot projects will be vital to test reliability, maintenance needs, and cybersecurity under practical conditions. Studies considering local climate, driving habits, and grid constraints will guide optimal sizing of PV arrays and batteries. Ensuring compatibility with evolving standards such as ISO 15118 for V2G and OCPP for charging management will help these stations fit into broader energy and transport networks. Making installations easier and more affordable is key to wider adoption. Using second-life EV batteries and integrating with home solar plus storage setups can cut costs. Plug-and-play hardware modules and automated software setups would simplify deployment and operation. As FPGA technology advances, becoming more powerful, energy efficient, and affordable, it will increasingly enable solar EV chargers that are reliable today and ready for tomorrow's innovations. To sum up, combining solar power with FPGA-based load management marks a major leap toward building EV charging systems that are efficient, dependable, and easy to scale. The unique capabilities of FPGAs such as high-speed parallel processing with precision control and flexibility enable optimal energy harvesting with seamless power flow coordination. As more renewable energy comes online and smart grid technology advances, these smart charging stations will become key players in balancing electricity demand extending battery life and improving the overall experience for users. With ongoing progress in FPGA technology, power electronics, and intelligent control methods we can expect these solar-powered, FPGA-controlled chargers to become foundational in creating sustainable and resilient transportation networks in the years ahead.

REFERENCES

1. C.-Y. Chung, J. Chynoweth, C. Qiu, C.-C. Chu, and R. Gadh, *Design of Fast Response Smart Electric Vehicle Charging Infrastructure*. Department of Mechanical and Aerospace Engineering, University of California, Los Angeles, USA, 2012.
2. K. Swamynathan, N. Mahalingam, A. Paramasivam, and S. Vijayalakshmi, PV based OFF grid charging station for E-vehicles using PWM and phase shift controlled interleaved three port converter. *SN Applied Sciences*, 5, 331, 2023.

3. R. W. Hamad, M. A. Qasim, and S. Raed, FPGA-based charging stations for electric vehicles: a review. *Journal of Modern Computing and Engineering Research*, 2024, 68–81, 2024.
4. A. Ul-Haq, C. Cecati, and E. A. Al-Ammar, Modeling of a photovoltaic-powered electric vehicle charging station with vehicle-to-grid implementation. *Energies*, 10(1), 4, 2016.
5. A. R. Singh, R. S. Kumar, K. R. Madhavi, F. Alsaif, M. Bajaj, and I. Zaitsev, Solar wireless electric vehicle charging system. *Journal of Information and Optimization Sciences,* 45, 829–838, 2024. doi:10.47974/JIOS-1594.
6. K. Elango and B. Sarala, FPGA-based photovoltaic efficient energy management with reversible off the grid power supply. *International Journal of Engineering Research & Technology*, 4(14), Paper ID: IJERTCONV4IS14007, 2018.
7. U. M. Damodarin, G. C. Cardarilli, L. Di Nunzio, M. Re, and S. Spanò, Smart electric vehicle charging management using reinforcement learning on FPGA platforms. *Sensors*, 25(8), 2585, 2025. doi: 10.3390/s25082585.
8. M. Ragulkumar, P. Manikandan, and G. K. D. Prasanna Venkatesan, FPGA based optimal charging in a solar powered robot. *IOSR Journal of VLSI and Signal Processing,* 4(3), 29–33, 2014. doi: 10.9790/4200 04322933.
9. R. Kuselan and V. Sundharajan, Investigation of integrated converter circuit and enriched error tolerant fuzzy logic controller based control approach for solar powered EV system. *Circuit World*, 49(2), 229–239, 2022. doi:10.1108/CW 12 20210317.
10. M. Sajid, T. Porselvi, and E. Abdul Sathar, FPGA-based multiple high-frequency electric vehicle charger for smart charging stations. *Protection and Control of Modern Power Systems*, 5(18), 2020.
11. R. Daisy Merina, R. Saravana Ram, and L. C. P. Micheal, FPGA implementation of intelligent battery management unit (i-BMU) based on dynamic power distribution technique for electric vehicles. *Analog Integrated Circuits and Signal Processing*, 123(42), 2025. doi: 10.1007/s10470-025-02370-8.
12. G. Mathesh and R. Saravanakumar, A novel intelligent controller-based power management system with instantaneous reference current in hybrid energy-fed electric vehicle. *Frontiers in Energy Research*, 11, 1130401, 2023. doi: 10.3389/fenrg.2023.1130401.
13. S. M. Shariff, M. S. Alam, F. Ahmad, Y. Rafat, M. S. J. Asghar, and S. Khan, System design and realization of a solar-powered electric vehicle charging station. *IEEE Systems Journal*, 14(2), 2748–2758, 2020. doi:10.1109/JSYST.2019.2931880.
14. L. Shanmugam, P. M. Balasubramaniam, A. H. Shnain, R. Abdulkader, M. Alharbi, D. A. Pustokhin, I. V. Pustokhina, S. Mohammed, and S. Sengan, Solar-powered LVDC nano-grid with smart FPGA-based residential power switching algorithm. *IEEE Access*, 11, 32001–32014, 2023. doi: 10.1109/ACCESS.2023.3240651.
15. N. Sharma, P. Sharma, D. Irwin, and P. Shenoy, Smart grid-based EV infrastructure with renewable integration. *Energies*, 10(1), 123, 2017.
16. S. Hussain, M. A. Ahmed, Y.-C. Kim, and Y.-C. Kim, Energy management for PV-based charging stations for electric vehicles. *Energies*, 14, 2021.
17. A. Randive, S. A. Ghadage, V. S. Gholap, and S. N. Maskare, Wireless electric vehicle charging system using solar energy. *International Journal for Research in Applied Science and Engineering Technology*, 12(7), 1015–1018, 2024.
18. U. Subramaniam, K. S. Reddy, D. Kaliyaperumal, V. Sailaja, P. Bhargavi, and S. Likhith, A MIMO-ANFIS-controlled solar-fuel-cell-based switched capacitor z-source converter for an off-board EV charger. *Energies*, 16(4), 1693, 2023.
19. A. Patidar, V. K. Dubey, and D. Sarvate, Design and development of solar based fast charger for e-vehicle using MPPT algorithm. *International Journal of Renewable Energy Research*, 14(2), 364–369, 2024.
20. D. Yan and M. Chengbin, Stochastic planning of electric vehicle charging station integrated with photovoltaic and battery systems. *International Journal of Electrical Power & Energy Systems*, **113**, 576–588, 2019.
21. U. Farooq, M. Jamil, M. A. Shah, and T. K. Saha, Wind–solar hybrid energy system: a review. In: *IEEE Power and Energy Society General Meeting (PESGM)*, Boston, 2016.
22. V. R. Paladi and S. Mahadeva Prasanna, A MIMO variable bit rate speech transmission for indoor wireless communication. In: *Proceedings of International Conference on Information and Communication Technologies and Applications (ICICTA) 2008*, Nagpur, 2008, pp. 1–6.

Integration of GPS-Based Real-Time Navigation with Smart Building Automation for Seamless User Experience

7

M. S. Nanditha, S. S. Radhikaa,
S. Subhiksha, and O. V. Gnana Swathika

7.1 INTRODUCTION

Smart buildings use key drivers like technology, comfort, and energy efficiency. These drivers connect with systems such as lighting and security. Case studies show real-world results from intelligent designs [1]. Rational design ideas by Durand focus on function and simple building systems. His approach supports smart, low-cost, and organized construction [2]. Smart energy systems collect data from sensors and adjust lighting and temperature based on user needs. Rule-based systems help save energy without reducing comfort [3]. Smart buildings use sensors, GPS, and controllers to track people, the environment, and building use. They send real-time data to control systems, which helps during emergencies and improves safety [4]. Internet of Things (IoT) and artificial intelligence manage lights, gates, and alerts based on where people are. These tools improve daily building tasks and support safe evacuation [5]. Building Information Modeling (BIM) connects systems like lighting and Heating, Ventilation, and Air Conditioning (HVAC), helps avoid design issues, and supports smart building operation [6]. Co-simulation platforms test how buildings, power grids, and tools work together, improving energy use and control across city areas [7]. GPS tracks tall buildings during wind or earthquakes,

DOI: 10.1201/9781003649700-7

improving safety and reducing costs [8]. Evacuation systems use mobile apps to guide people out safely based on personal needs and sensor data. Smart control systems manage A/C and lights using Bluetooth signals and machine learning. They save energy and help buildings reach low or zero emissions.

7.2 SUSTAINABILITY AND ENERGY-EFFICIENT BENEFITS OR COMPARATIVE ANALYSIS WITH EXISTING SMART SYSTEM

Yigitcanlar et al. performed bibliometric analysis to investigate trends in research on sustainability in smart cities. They examined 1,291 journal papers between 1991 and 2020 based on the Scopus database and observed a significant rise in publications post-2013. The work identifies prominent areas of research, including energy efficiency, smart governance, integration of information and communication technologies (ICT), urban resilience, and sustainable mobility. It indicates that China, the USA, the UK, Italy, and India are the leading contributor nations. The study also discovers that sustainability in smart cities is also a very interdisciplinary area that includes urban planning, environmental science, and engineering. It indicates an increasing application of technologies such as IoT, artificial intelligence (AI), and big data in order to facilitate sustainable development goals (SDGs). The authors hold the view that future advances in smart cities rely on effective cooperation and holistic growth that takes into account both technology and sustainability [1]. These are commonly used in smart cities, hospitals, and military systems. Since the individual systems within an SoS operate independently yet still communicate, they may interact unpredictably and thus become more challenging to secure. The authors focus on one category of threat known as cascading attacks, in which a problem propagates through the system and renders more severe problems. To address this, they employ a modeling process named Model-Driven Engineering (MDE) and develop a tool and a language known as SoSSecML, and a simulation environment based on Multi-Agent Systems. They try out their approach in a real smart building—Adelaide Health and Medical School—to test its effectiveness. The approach identifies covert attacks and enhances system design prior to it being constructed. The authors hope that this tool can make smart systems more secure and intend to enhance it in the future by adding factors such as trust and safety [2]. The system employs the Constrained Application Protocol (CoAP) protocol to provide low-power communication and combines Datagram Transport Layer Security (DTLS) with SHA-256 to provide safe data transmission. Simulation outcomes indicate that CoAP decreases energy consumption by 30.86% with respect to MQTT. The research identifies the way IoT and big data assist in monitoring buildings, managing systems such as HVAC, and enhancing comfort and security. With the Contiki Cooja simulator, the authors show that their suggested design is efficient and secure. The authors consider that IoT will be an important part in the future of intelligent buildings [3]. This chapter presents an energy-efficient smart building design using modern IoT technologies. It uses CoAP for lightweight communication and employs DTLS coupled with SHA-256 for secure data transmission. Testing is done through the Contiki Cooja simulator, and results show an approximately 30.86% reduction in energy consumption as compared to using MQTT. The architecture provides real-time supervision of building systems like HVAC, lighting, and occupancy for better comfort and security. The study further highlights the use of cloud computing and big data for real-time managing of sensors and data stream for smart decision-making. Overall, this study illustrates the versatility of IoT in enabling buildings to become more energy-efficient, secure, and responsive [4]. This work develops a smart building energy management system (SBEMS) using a multi-agent control framework with intelligent optimization. It seeks to optimize energy consumption and comfort levels indoors by considering user preferences and the outdoor climate. The system relies on evolutionary algorithms Multi-Objective Genetic Algorithm (MOGA) and Hybrid Multi-Objective Genetic Algorithm (HMOGA) to optimize temperature, lighting, and air quality relative to the comfort indices. A hybrid approach achieves 31.6% energy savings, with an 8.1% improvement in user comfort. The system models a building equipped with solar panels and wind turbines in islanded mode. It has advanced controllers that manage actuators and energy distribution. In summary, the proposed method fosters advanced energy-efficient control, comfort, and sustainability in smart buildings through the use of renewable energy, sophisticated control algorithms, and HMOGA [5]. The context-aware IoT approach enhances energy efficiency while improving occupant comfort in smart building management systems. It proposes a system that collects real-time sensor data from context-aware engines. This engine makes decisions based on environmental data, user activities, and device conditions. A machine learning-based layered architecture for sensing, processing, and applications is provided in the chapter. It automates energy management and controls building features such as lighting and HVAC systems based on predicted energy consumption. While ensuring comfort, the system has minimized unnecessary energy consumption. Simulation results significantly improved energy savings. This approach incorporates real-time data and prediction models for the smart building control system [6]. In order to lower energy consumption and enhance user comfort, the Intelligent Context-Awareness Building Energy Management System, or ICA-BEMS, employs both passive and active energy-saving strategies. To comprehend complex building environments, ICA-BEMS combines context awareness and smart building ontology. It uses intelligent reasoning and real-time sensor data monitoring to identify energy waste and make the right choices. As building conditions and user behavior change, the system effectively schedules appliance usage. Protégé ontology tools are used in its implementation, and a simulator is used for testing. The energy consumption has decreased by 40%, according to the results. In dynamic smart building environments, ICA-BEMS shows how context-aware systems and hybrid energy-saving techniques can be combined to effectively increase energy efficiency while preserving comfort [7]. The application of a SBEMS that combines IoT and machine learning to increase energy efficiency is covered in this chapter. To keep an eye on variables like temperature, humidity, and occupancy, the system gathers data in real time from sensors. In order to identify trends and make wise choices for managing the HVAC, lighting, and other appliances, machine learning algorithms examine this data. This method preserves indoor comfort while cutting down on wasteful energy use. The three main tactics highlighted in the chapter are automation, predictive control, and user behavior learning. The SBEMS successfully reduces energy waste and improves building performance, according to the results. All things considered, the study shows how IoT and machine learning can be combined to create smarter, more effective building operations for sustainable energy use [8]. This study offers a rule-based system that was created through data mining and expert knowledge to identify energy inefficiencies in smart buildings. In order to detect irregularities in energy consumption, it presents Energy Efficiency Indicators (EEIs). Sensors around the building gather data, and the system uses decision rules based on past data to identify and fix inefficient behaviors. The BlueNet building is used to test the framework, and the results indicate that it can optimize energy use while preserving user comfort. In keeping with the objectives of smart cities, the system seeks to lower energy use and environmental impact. Future plans call for incorporating the model into Eugene, a commercial energy management program that enables real-time building energy performance monitoring and control. The study suggests a data-driven approach that makes use of fault detection and diagnostic (FDD) systems to improve energy efficiency in smart buildings. In order to identify and fix inefficiencies in HVAC operations, it focuses on integrating machine learning models with building management systems. The study offers a framework for locating errors and improving performance using real-time data. This method uses rule-based diagnostics and classification algorithms to assess system behavior. The findings demonstrate that integrating data analytics and FDD lowers operating costs and increases energy efficiency. The significance of precise sensor data and system modeling is also emphasized in the chapter. Overall, by automating fault detection and improving energy-saving techniques, the framework facilitates better decision-making in smart buildings. This study examines how real-world occupancy data can enhance building comfort and energy efficiency. Since HVAC systems consume the most energy, it focuses on them. Using cameras and deep learning, the researchers gather occupancy data. They then experiment with varying time intervals (5–60 minutes) to regulate the HVAC system. They discover that while longer intervals perform better in stable settings, shorter intervals save more energy in crowded areas. The system uses programs like EnergyPlus and OpenStudio for simulations and operates on a real multi-zone office. According to the study, occupancy-centric control, or occupancy-centric control (OCC), can help save energy while maintaining occupant comfort. It promotes more practical uses of OCC in smart buildings and offers building managers better ways to use it. This chapter describes how the IoT contributes to the intelligence and efficiency of buildings. It demonstrates how sensors, cloud computing, and data analysis are used in smart buildings to regulate energy use, lighting, and temperature. According to the chapter, IoT enables devices to gather and exchange data by connecting them to one another. This saves energy and improves building comfort. It also demonstrates how IoT is applied at various phases of building operation and design. It does, however, note that further study is required as the full application of IoT in buildings continues to expand. In order to improve the way IoT is used in smart buildings and achieve better outcomes in the future, the chapter encourages experts from various fields to collaborate. This chapter discusses how energy management systems and the IoT are used in smart buildings to reduce energy consumption and enhance comfort. It describes how sensors, smart meters, and other gadgets gather information about occupancy, temperature, and lighting. Control systems use this data to inform their energy-saving decisions. The study also emphasizes the significance of data analysis and device communication in building intelligence. It states that machine learning can be used to improve system control and forecast energy requirements. The study discusses the advantages and difficulties of the various technologies and approaches utilized in smart buildings. In general, it claims that energy-saving technologies and smart buildings with IoT lower energy expenses and promote a more sustainable future. The text talks about how smart technologies may assist smart buildings save energy. It focuses on how data systems and the IoT aid in enhanced energy management. It describes how sensors and smart devices gather information and use it to inform decisions that lower energy consumption without compromising occupant comfort. The study emphasizes how ICT help people connect devices and make informed decisions. Additionally, it displays European smart building examples and their performance. According to the chapter, more research is required to improve the functionality of smart buildings. It advocates for a greener future through the use of digital tools.

7.3 USE CASES AND APPLICATIONS IN SMART BUILDING

Smart building systems manage access through digital controls and automation. Stamatescu et al. explain that modern buildings use systems like BACnet to link access points, sensors, and user data. These systems allow entry only when the rules match—for example, opening a door when a person is nearby. Using ESP32 with NEO-6M GPS, the system checks if a person or vehicle reaches a set location. If true, the ESP32 sends a signal to open the gate or door. This limits access to secure zones and improves building safety. The chapter also points out the risk of weak spots in access systems and highlights the need for careful setup to avoid attacks. Smart buildings use IoT-based systems to improve safety, security, and automation. ESP32, when used with the NEO-6M GPS module, supports location-aware features like real-time occupant tracking and automated access control. The system collects data like position and activity, which helps the building decide when to open gates, send emergency alerts, or trigger safety actions. This setup lowers manual work and improves response speed, making the building smarter and more responsive to real-world situations. Smart buildings use technologies like sensors, controllers, and automation systems to improve energy efficiency, comfort, and safety. These systems help lower running costs and improve user experience. The study shows that smart buildings focus on environmental sustainability and use networked devices to collect and respond to real-time data. As building intelligence grows, it plays an important role in modern urban development, especially in countries like China where rapid growth pushes the need for efficient infrastructure. Smart buildings use the IoT to improve automation, comfort, and energy use. IoT devices like sensors and actuators help smart buildings collect data, monitor conditions, and control systems in real time. These technologies support applications such as lighting control, HVAC optimization, safety monitoring, and remote access. The study shows that IoT also helps connect different systems within a building, leading to better efficiency and user experience. Using IoT in buildings helps create environments that are adaptive, responsive, and more sustainable. Smart buildings use modern technologies to save energy and improve how buildings work. Smart HVAC systems use sensors and software to control heating and cooling based on room use, air quality, and weather. They also find problems early and help fix them before energy is wasted. Smart lighting uses motion and light sensors to turn lights off or dim them when rooms are empty or bright, and wireless controls make setup easier. Plug load controls like smart plugs and power strips stop power to devices when they are not in use, helping save electricity in places like offices and kitchens. Smart window shading controls sunlight and heat, using sensors or special glass to reduce cooling needs while letting in natural light. Energy management systems collect data and help building operators watch and control systems using dashboards or cloud tools. Some systems also predict weather and make changes before problems start. People in smart buildings use apps or screens to adjust lights, see energy use, and follow alerts. Smart energy systems like solar panels and batteries connect to the grid through smart inverters, which help manage power flow and respond to energy needs. All these technologies work together to lower energy use and improve comfort in buildings. Modern cities need tools to study how buildings, energy systems, and electric grids work together. A distributed co-simulation platform helps test different models at the same time across computers in different places. This platform connects building models, power systems, and communication networks so that they run together during a test. It uses a central server to link all parts and manage the time and data between them. Different software tools send and receive messages to keep the test in sync. This setup lets users try out smart energy systems, like buildings that change their energy use based on grid needs or prices. It also helps study large systems, like entire city districts, while keeping the models flexible and reusable. By using this approach, cities and researchers find better ways to save energy, improve power use, and support smart grid functions. Modern power systems need smart ways to handle fast changes in energy supply and demand. A smart energy controller uses real-time data and AI to make better decisions about how to use energy. This system works with microgrids, which are small, local energy systems that include solar power, batteries, and other energy sources. The controller uses sensors, communication networks, and learning tools to decide when to store energy, when to use it, and when to share it with the main power grid. It also helps keep the system stable when there are faults or blackouts. This kind of system supports clean energy, reduces energy loss, and makes energy use more efficient. It also helps lower costs and improve reliability by adjusting to what is happening in real time. Compact LoRa devices help improve smart building systems. Many smart devices in buildings face size and antenna placement limits. A small LoRa-based device is designed to use low power and fit in small spaces. It includes a built-in antenna and supports both sensing and control. LoRa works better than other wireless options like WiFi or Z-Wave because it uses less power and works over longer distances. Tests show that the device gives strong signals even up to 4.2 km in open areas. The antenna design supports this performance. Simulations and real tests confirm that the system is reliable for smart building use. A smart solar energy system manages how energy is used in homes. It uses IoT sensors and solar panels to track how much energy is made and used. The system decides when to use solar power, when to store it, and when to send extra power back to the grid. It also tries to reduce electricity bills and avoid energy waste. The data is saved in the cloud, and users can control the system from anywhere. The setup helps support green energy and better energy use in homes. A rule-based model helps smart buildings take better control of energy use and comfort.

The system follows clear if-then rules to manage things like lights, heating, and cooling. It uses inputs from sensors, weather data, and user settings to decide what actions to take. The model also detects faults and responds quickly when problems happen. It works well with different building types and does not need full system changes when updated. The platform is flexible, easy to understand, and saves energy while keeping the building comfortable. Smart buildings use IoT devices like sensors, smartphones, and WiFi access points to track how many people are in a room or area. The system helps manage energy by adjusting heating, cooling, and lighting based on occupancy. Different methods, like camera-based counting or signal tracking from mobile devices, help detect presence. Using data from more than one source improves accuracy without needing extra hardware. These methods support energy saving, comfort, and safety. The chapter compares techniques based on accuracy, cost, and privacy, and suggests better ways to combine data using tools already available in buildings. A microservice-based platform helps smart buildings manage data from sensors in a flexible and efficient way. The system uses the Jolie programming language to connect devices, collect data, and control building functions like temperature, light, and humidity. Small devices send information using Bluetooth, and the system processes it using both Jolie and Java. Data is stored in files and analyzed to improve comfort and save energy. The setup runs on Raspberry Pi and works in real environments like student rooms and university labs. The model tracks how many people are in a room, when they enter or leave, and uses this to adjust conditions. It also supports future plans to build a larger network using more sensors and protocols like ZigBee and Z-Wave. The platform allows easy updates and connections for different types of devices, which makes it suitable for wider use in smart buildings. Cyber-physical systems help cities and buildings work smarter by connecting the physical and digital worlds. These systems sense changes in the environment and react to them through real-time decisions. They help manage energy, improve comfort, and support safety in smart cities. The chapter explains that self-aware systems follow a loop: they monitor data, analyze it, plan what to do, and carry out actions. Important features include self-adaptation, self-organization, and self-optimization. These systems work better when combined with cloud computing, which stores and processes large data streams from sensors. A service-oriented middleware collects sensor data, manages devices, and makes decisions based on changing conditions. This model allows cities to run services like traffic control, waste management, or air monitoring more efficiently. The design supports easy changes, local control, and energy saving through smart decisions, making the system flexible and ready for future use. Smart buildings use automated systems to manage energy, lighting, heating, and security. The chapter gives an overview of different automation systems and how they support energy saving. It compares systems like KNX, BACnet, LonWorks, and others used in building automation. These systems connect sensors, controllers, and devices to make real-time decisions. They also help reduce energy costs and improve comfort by adjusting settings based on occupancy or time of day. Each system has its own features, setup process, and level of openness. The study explains the strengths and weaknesses of each one and suggests which systems fit better for different building types. Open protocols and flexible designs allow better communication between devices and help create smarter, more energy-efficient buildings.

7.4 INTEGRATION OF REAL-TIME NAVIGATION WITH SMART INFRASTRUCTURE

To safely guide people during emergencies, we use indoor localization technologies like WiFi, RFID, and iBeacons. iBeacons work best because they are cheap, accurate, and smartphones can read them easily. Phones detect beacon signals and send location info to a central system. They also use built-in sensors to track user activity like walking or running, helping to spot panic. Extra sensors like smoke and heat detectors give more safety data. All this information helps the system guide people and manage the evacuation better. The proposed method lets users control electric appliances by pointing and clicking with a handheld device. It uses ultra-wideband (UWB) positioning and Inertial Measurement Unit (IMU) sensors to track the user's indoor location and direction. This data is sent to a virtual model of the building (BIM) to identify the targeted appliance. A simple click then switches the appliance on or off through an IoT network. The approach is more intuitive and reliable than voice or gesture controls and shows good results in tests. Its accuracy depends on precise indoor tracking and proper placement of devices. The IoT is becoming important in making buildings smarter, more efficient, and user-friendly. It connects everyday objects like lights, doors, and sensors to a network, allowing them to collect and share data. This helps buildings run smoothly, save energy, and improve comfort. However, IoT is still growing, and its use in construction and building operations is limited. More research is needed to solve challenges and fully integrate IoT into smart buildings. Experts from different fields must work together to unlock its full potential. A smart navigation wearable device helps blind and visually impaired people walk safely in a smart building or any other smart infrastructure. It has sensors on a headband and a handband to detect obstacles around and on the ground. The system uses a Raspberry Pi and Arduino to process sensor data in real time. It gives voice and vibration alerts to guide the user. A fuzzy logic controller checks the safety of the path using obstacle distance and walking speed. The device is lightweight, easy to wear, and runs on a rechargeable battery. Frankfurt Airport uses smart building technology to help firefighters during emergencies. A new system gives firefighters digital maps and real-time location using indoor positioning tools like RFID, UWB, and WiFi. This system replaces paper maps and helps rescuers find the fastest and safest path. It also shows important building details like gas lines or electric panels. The system uses BIM data to create navigation routes on mobile devices. It is tested in a university and will be used in real situations at the airport. The IoT helps devices like sensors, GPS, and RFID connect and share information in smart buildings. Indoor positioning is hard because GPS doesn't work well indoors, so systems like WiFi, Bluetooth, and RFID are used instead. The system needs to handle both moving and fixed devices, and work well in places with signal problems like interference and reflection. IoT location algorithms must be fast, low-cost, secure, and able to adjust in real time. For this, hybrid algorithms that combine different methods are best. Future work focuses on a new smart system that understands the location and context of people, objects, and services in a building. Smart buildings need better ways to help firefighters quickly rescue people during fires. This study introduces a low-cost emergency system where trapped people scan a QR code on doors to send their location to firefighters. Since smoke and darkness make rescues harder, knowing exact locations helps save lives faster. Many buildings still lack strong rescue tools, and current indoor positioning technologies have limits. This QR-based system is tested with firefighters and found to reduce search times and improve rescue success. It is also cheaper and easier to use than high-tech AI systems. Many researchers study smart buildings to improve energy efficiency and comfort. The IoT helps by connecting devices like lights, doors, and sensors so they can share data and make buildings smarter. However, current systems are not fully connected, and applying IoT in buildings is still a work in progress. IoT combines different technologies like wireless sensors, big data, and cloud computing. While most IoT research focuses on tech, more work is needed from civil and building experts to make it work well in real buildings. Real-time navigation in smart buildings helps people move easily inside large and complex spaces. It uses sensors, indoor maps, and technologies like WiFi, Bluetooth, or RFID to track and guide users. This system shows the shortest route to rooms, exits, or important spots on mobile devices. It is very useful during emergencies by helping people and rescuers find safe paths quickly. Smart buildings use this navigation to improve safety, save time, and make the building more user-friendly. Similarly in smart buildings, it also helps visitors or staff find their way inside large or complex spaces, like hospitals, airports, or malls. It gives step-by-step directions on a mobile device to reach specific rooms or services. This saves time and reduces confusion, especially for first-time visitors. It also helps people with disabilities by suggesting accessible routes. The system updates instantly if a path is blocked or changed. In buildings with big parking areas, real-time navigation helps drivers find empty spots without driving around for long. Sensors in the parking lot detect which spots are free or taken. This information is sent to a mobile app or digital display at the entrance. Drivers can follow signs or use the app to go directly to an available space. This saves fuel, reduces air pollution, and avoids traffic jams inside the parking zone. The system also remembers where you parked your car and shows you the shortest route to find it later. It makes parking fast, stress-free, and more organized in smart buildings. In smart buildings, real-time navigation helps indoor delivery robots move through hallways and floors to deliver food, parcels, or documents. These robots use indoor maps and real-time data to avoid obstacles like people, furniture, or closed doors. They take the fastest route, use elevators when needed, and can even notify the recipient when they arrive. This system makes delivery inside large offices, hospitals, or hotels quick and contactless. It also reduces human effort and improves convenience for everyone inside the building.

7.5 CASE STUDY

7.5.1 Real-Time Navigation Using ESP 32 and Neo-6M GPS Module

7.5.1.1 Introduction

This project involves integrating a real-time navigation system with a smart building automation system. The goal of this project is to help the user experience seamless entry by triggering the appliances when they come closer to the building. It enhances security through location-based access and also supports energy saving by turning off the appliances when the user exits.

7.5.1.2 Objective

There are three main objectives of this project. Firstly, it receives and interprets real-time GPS data. Then, using a communication system, it communicates the directions through audio instruction. And finally, it prepares a seamless handover to smart building automation (Figure 7.1).

7.5.1.3 Components used

1. ESP32 Microcontroller
2. NEO6M GPS Module
3. Li-ion Battery + TP4056
4. PAM8403 Audio Amplifier
5. Breadboard and jumper wires

7.5.1.4 Block diagram

7.5.2 GPS Data Acquisition

The foundation of this project is to connect the NEO-6M GPS module to the ESP32 microcontroller to obtain real-time location data. To effectively process the data, the TinyGPS++ library was integrated to get accurate values of latitude and longitude. The values of latitude and longitude will be continuously monitored and displayed in serial output, which ensures the user's movement with precision.

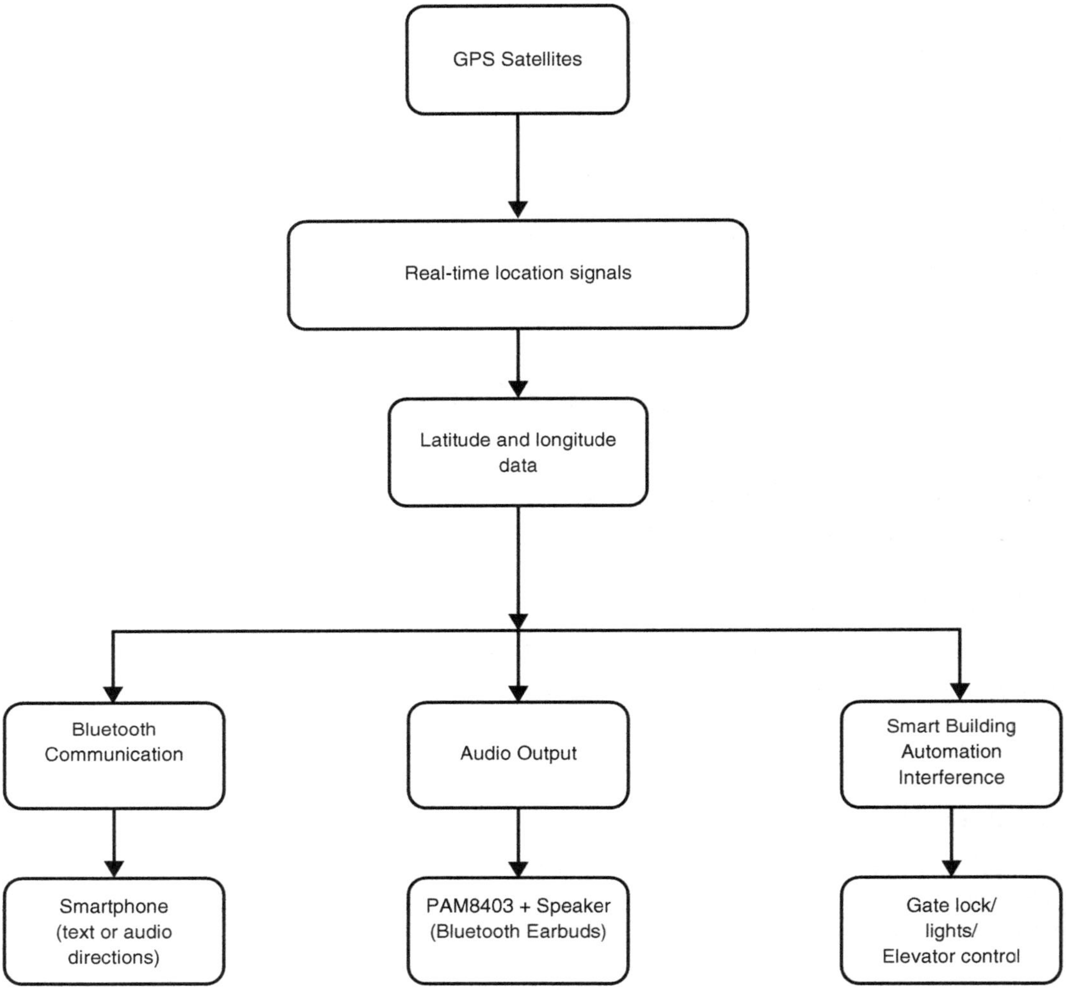

FIGURE 7.1 Methodology.

7.5.3 Real-Time Navigation Output

The ESP32 microcontroller utilizes the BluetoothSerial library to convert text-based directional messages to an audio-based Bluetooth device. Messages like "Turn right and walk through 100 meters" are used to guide the user seamlessly through outdoor environments to the smart building. This mechanism helps the user avoid the constant visual reference to the smartphone or display.

7.5.4 Audio System Preparation

To elevate the user experience through hands-free interaction, many audio output methods were explored, which includes pre-recorded voice directions stored in SPIFFS. This enables Bluetooth audio streaming via A2DP for wireless earphones and implementing dynamic voice instructions using online Text-to-Speech (TTS) services via WiFi module.

7.5.5 Smart Building Automation Context

The system predicts a smooth integration between outdoor navigation and indoor building automation. When the user arrives, detected by GPS proximity, the system initiates automated actions such as lighting the lights, opening the doors, or calling the elevators. These features provide a seamless and context-sensitive experience that aligns with the vision of smart and responsive buildings.

7.5.5.1 Current progress

The ESP32 and NEO-6M module have successfully acquired GPS data, guaranteeing precise real-time location tracking. The implementation of a Bluetooth-based real-time direction messaging system enables the wireless transmission of navigation instructions to a device that is connected. With the goal of improving user interaction through spoken navigation cues, the audio output design is presently under development. In order to provide a smooth and intelligent user experience, the next stage of integration will concentrate on integrating smart building automation, starting with automated entry point systems.

7.5.5.2 Challenges faced

GPS accuracy often drops near tall buildings or in places where signals get blocked. The ESP32 also has limited memory, which makes it hard to run both dynamic TTS and advanced automation at the same time. As of now, there's no indoor positioning system added, since GPS doesn't work well inside buildings.

7.6 CONCLUSION

Our project uses the ESP32 microcontroller with the NEO-6M GPS module to track the real-time location of people inside or around smart buildings. The system gives audio directions using a speaker connected through the PAM8403 module. This helps people move easily through complex areas, find exits, or follow safe paths during emergencies. It is especially useful in places like airports, offices, and large commercial buildings where navigation can be confusing.

In smart buildings, automation plays a big role. Our project supports automation by allowing the building to react based on where people are. For example, when someone enters or leaves a room, the system can turn lights or A/C on or off to save energy. This improves sustainability and reduces electricity use, which is one of the major goals of green smart buildings.

During emergencies like fires or gas leaks, the real-time location data can be sent to a control system to alert rescue teams or guide people to safety. This improves safety and supports faster rescue operations. Smart buildings also use this type of system for managing services like cleaning, parking, and security.

By using affordable and simple IoT components, our project shows that it is possible to build smart features without expensive technology. It connects well with current smart building strategies and offers real-life benefits. Overall, our navigation system adds value to smart buildings by making them safer, smarter, and more energy-efficient.

REFERENCES

1. A. De Paola, G. Lo Re, M. Morana, M. Ortolani (2017). Adaptive context-aware learning for smart building energy management. *Journal of Ambient Intelligence and Humanized Computing*, 8, 715–730. doi:10.1007/s12652-016-0418-4.
2. J. N. L. Durand (2000). Jean Nicolas Louis Durand: Theory of Architecture. https://nortech.oulu.fi/files/Jean_Nicolas_Louis_DW.pdf.
3. A. Sharma, R. Dwivedi, M. Kaushik et al. (2021). Green smart building: requisites, architecture, challenges and use cases. In: A. Sharma, R. Dwivedi, and M. Kaushik (Eds.), *Green Smart Building—Requisites, Architecture, Challenges and Use Cases* (pp. 25–72). Wiley.
4. J. King and C. Perry (2017). *Smart Buildings: Using Smart Technology to Save Energy in Existing Buildings*. American Council for an Energy-Efficient Economy.
5. S. I. Lopes, F. Pereira, J. M. N. Vieira, N. B. Carvalho, A. Curado (2019). Design of compact LoRa devices for smart building applications. In: J. L. Afonso et al. (Eds.), *GreeNets 2018: Lecture Notes of the Institute for Computer Sciences, Social Informatics and Telecommunications Engineering* (Vol. 269, pp. 1–12). Springer.
6. M. Jia, A. Komeily, R. S. Srinivasan (Eds.) (2019). Adopting Internet of Things for the development of smart buildings: A review of enabling technologies and applications. *Automation in Construction*, 101, 111–126. https://www.sciencedirect.com/science/article/abs/pii/S0926580518307064.
7. Z. Turgut, G. Zeynep Gurkas Aydin, A. Sertbas (2016). Indoor localization techniques for smart building environment. *Procedia Computer Science*, 83, 1176–1181. https://www.sciencedirect.com/science/article/pii/S1877050916302757.
8. K. M. Rashid, J. Louis, K. K. Fiawoyife (2016). Wireless electric appliance control for smart buildings using indoor location tracking and BIM-based virtual environments. *Procedia Computer Science,* 83, 480–487.

Cognitive Security for Smart Buildings

8

A Review of EEG-Based Multimodal Authentication System

Deepa Rakesh Kumar, Kumar Sairaj, Daksha Raghunath Bordekar, Maryamah, and O. V. Gnana Swathika

8.1 INTRODUCTION

Recent studies into Brain-Computer Interface (BCI) or Brain-Machine Interface (BMI) for both medical and non-medical systems raise significant issues regarding security and privacy. Attacking BCI systems could result in the interception and alteration of neural data and also malware attacks, leading to misuse of previous neural data. As electroencephalogram (EEG)-based devices can expose raw brain signals, it could also allow attackers to exploit identity spoofing, accidental and unauthorized control of prosthetics, or data extraction of sensitive information such as passwords and private information; these systems are subject to numerous security and privacy threats. Some of the proposed countermeasures include establishing secure design criteria for transceivers or encrypted elements and enhanced encryption protocols, such as functional, homomorphic, and differential privacy frameworks, along with developments in anonymization tools, such as the BCI Anonymizer tool, in order to clean EEG signals of inappropriate or unintended information. Joint measures, such as a greater enhancement of operating system-level permissions and awareness training for users about brain privacy, will ensure the ability to have sustained protection of BCI systems [1]. The paper [2] discusses how brainwave signals picked up through EEG provide a unique, highly secure identity

DOI: 10.1201/9781003649700-8

verification method. The authors focus on three basic ways this can happen: (1) by performing mental tasks, (2) observing brain activity while in a resting state, or (3) examining how brainwaves respond to specific external stimuli. The study also discusses ways to combine EEG with cryptography to protect the biometric information and to provide the ability to reset the signals should it be necessary for immediate security, and in essence provide another layer of security. However, they acknowledge challenges in acquiring EEG signals that can change due to emotional state or environmental conditions, and they observe how current EEG components are often bulky and uncomfortable. The authors emphasize that in order for EEG to become feasible for widespread use, there needs to be more sophisticated hardware than is currently available, more clearly defined standards for data acquisition and application for various uses, and more adaptive machine learning (ML) approaches to improve the EEG data reliability. The authors noticed that while combining EEG collections with other biometric types, additional reliability and legitimacy can be achieved. Overall, the authors are optimistic about the use of EEG in security access but note, honestly, a great deal of work is needed before EEG will be a preferred verification option for day-to-day access. Chaurasia et al. [3] discuss the development of NeuroIDBench, which is an open-source benchmark framework that enables standardized methodology for EEG-based authentication research to address issues such as disparate protocols, small datasets, and poor reproducibility. NeuroIDBench encompasses nine available EEG datasets, an extensible preprocessing pipeline, and it supports seven types of classifiers, including shallow models such as Random Forest, Support Vector Machine (SVM), and deep models such as Twin Neural Networks (NNs). NeuroIDBench also enables performance evaluation under multiple adversary models—"known" when the attacker was identified vs "unknown" when the attacker was not identified, and can evaluate single-session as well as multi-session scenarios. Single-session testing typically offers data from one session, often resulting in overly optimistic accuracy estimates. However multi-session operational scenarios introduce time-based changes, such as different days and times, different mental states, noise, and hardware variance. These time-based variables create scenarios that are much closer to realism while also testing the models' ability to generalize across time—a crucial aspect of actual deployment in authentication-type scenarios, which is the goal of this review. The authors achieved a 37.6% decrease in Equal Error Rate (EER) for unknown attacker scenarios and noted that Power Spectral Density (PSD) and the Auto-Regressive model of order 1 (AR1) are feature-rich datasets best for shallow classifiers. They also established that epoch rejection had little impact on performance. Further, the results indicated that when data was limited, traditional classifiers, particularly in multi-session scenarios, performed better than deep learning classifiers. The authors also emphasized best practices, including fairness, sampling, transparency, and responsible use. NeuroIDBench presents a flexible and extensible platform that will facilitate better comparisons and generate more rapid progress to realize practical, ethical neuro-authentication systems. Isa et al. [4] examine physiological responses of the alpha rhythm associated with the brain's relaxed or calm state, induced by minimal stimulus: the command "eyes open" or "eyes closed". The authors recorded EEGs of two healthy young adults and found alpha wave activity was noticeably enhanced in trials with "eyes closed" and diminished with the "eyes open" trial, or mental alertness. The alpha wave consistency suggests alpha is a reliable measure when assessing relaxation. The authors developed an original cost/benefit EEG system to record signals from participants and found that those people that had stable alpha activity were composed and relaxed; these findings have potential future uses in rehabilitation, with brain activity, such as alpha rhythm, for control of FES (Functional Electrical Stimulation) in rehabilitation, and brain signals can potentially assist a stroke patient in their recovery of movement. Overall, these findings will enable researchers to measure physiological activity in neuro-rehabilitation, leading to smarter, more interactive tools for therapy. Priyadarshani et al. [5] elaborate on the topic of how emotion recognition has recently grown to be very important in burgeoning areas like healthcare, affective computing, and human–computer interaction. However, emotional interpretation based on facial expressions and textual interpretations based on emotion words usually lack validity and reliability, as communication is subjective and inaccurate. EEG provides a more objective and real-time method of measurement for detecting internal emotional states. The authors provided another emotion recognition opportunity using EEG samples and facial expression by demonstrating emotion recognition accuracy of 86.5%. The paper also talks about deep learning for emotion recognition, which can enhance the model adaptability and accuracy. The present study builds on previous work by using different deep learning models (Long Short-Term Memory (LSTM) and Gated Recurrent Units (GRUs)) using EEG data and readily available emotion datasets. Researchers utilized Butterworth filters, Fast Fourier Transform (FFT) methods, Principal Component Analysis (PCA), and PSD to preprocess and link EEG features to emotions. Following a review of emotion datasets, the researchers accessed an EEG emotion dataset from Kaggle. LSTM and GRU networks were effective by capturing the non-stationarity of EEG features, time components, and temporal patterns of EEG signals. Among the presented research, the LSTM was a very good classifier producing an accuracy of 97% using cross-validation. The developed LSTM and GRU models beat SVM and Random Forest from previous studies and produced high accuracy on a difficult emotion recognition problem. The great performance of the LSTM model was also valid against confusion matrices, ROC curves, and Area Under the Curve (AUC). The LSTM model demonstrated high true-positive rates for all emotional categories. The training-validating plots confirmed that the learning was stable and that overfitting was reduced. Altogether, this work demonstrates the potential of deep learning for decoding emotions from EEG signals. Compared to traditional approaches for emotion recognition, this represents a more robust, scalable, and ultimately more useful approach. This work also sheds light on applications in mental health, human–computer interaction, and adaptive systems. Finally, future research should explore larger datasets and more advanced feature extraction techniques. Recent studies in the field of EEG-based emotion recognition have led to the advancement of predictive models due to deep learning for sentiment analysis. In this study there was a proposed deep learning framework using EEG data recorded from a Muse headband, a non-invasive device with four electrodes positioned on the scalp. EEG signals underwent preprocessing and were analyzed using statistical, entropy-based, frequency-based features, and log-covariance matrices to represent brainwave interactions. Information Gain, Symmetrical Uncertainty, and Evolutionary Algorithms were feature selection approaches used to reduce noise and improve input data quality. Two deep learning models—LSTM and Multilayer Perceptron–artificial neural network (MLP-ANN)—were implemented and evaluated using both 10-fold cross-validation and train-test splits. The LSTM model showed exceptional performance with its time-series processing capabilities, outperforming the MLP-ANN with an accuracy of 97.19%, and a great F1-score, showing its effectiveness in identifying emotional states. The paper [6] demonstrates the number of opportunities that deep learning presents in real-time EEG emotion classification, for applications within mental health surveillance, user experience studies, and emotion-aware human–computer interfaces. The study [7] investigates how human emotions can be easily classified by analyzing brainwave data obtained from EEG signals using FFT and Convolutional Neural Networks (CNN). Multiple brain activities were recorded for emotions ranging from happy to sad at multiple times. The EEG signals were processed using FFT to extract the five major types of brainwaves: delta, theta, alpha, beta, and gamma. These frequencies were visualized into images using Python Spyder, then converted into grayscale, squared, and labeled "0" for normal and "1" for high activity. The preprocessed images were used in the CNN model in Jupyter Notebook consisting of 10 layers, including convolutional, activation, pooling, flattening, and dense layers, with binary cross-entropy loss. The model was able to predict whether brainwave activity was high or normal, and emotional states were collected based on wave combinations that are low alpha, delta, and theta with high beta and gamma indicating emotions.

Emotion recognition using EEG signals is crucial in Artificial Intelligence (AI) and has significant implications for emotional care, human–computer interaction, and multimedia content recommendation.

While many studies were conducted that reviewed EEG-based emotion recognition, the studies either have disconnected content or only review certain techniques adopted in this field without seeing the whole of the entire technical route. The paper [8] thus provides a review from the perspective of researchers trying to take the first step on this topic. This paper reviews the more recent representative studies in EEG-based emotion recognition research and provides a tutorial for researchers so that they can start fresh. This presents the scientific basis of EEG-based emotion recognition both psychologically and physiologically. Also this classifies the reviewed articles by listing several new technical routes and introducing their theoretical basis and research motivations, which ought to help any reader to understand why those techniques were adopted and used to study the problems. Finally, there are also current challenges addressed in this chapter that inform them about decisions regarding potential future research. May et al. [9] present an elaborate architecture for allowing secure transactions in Extended Reality (XR) environments like Virtual Reality (VR), or Augmented Reality (AR) using behavioral biometric authentication. The main technical innovation in the study is that it focuses on passively measuring an individual's inherent behavioral characteristics while using the XR system. The behavioral data is converted into biometric tokens that are to be matched to user profiles that can contain content preferences and payment credentials. When a transaction is requested, some biometric data is collected again and verified before approving the transaction. The system has entirely removed the use for typical authentication methods like password entry, which helped in maintaining an engaging experience while users transact. Despite these advantages, the technical and ethical challenges need to be overcome first. Behavioral biometrics are always subject to variation over time and context influenced by levels of fatigue, illness, emotional and mental state, or environmental conditions, which raises questions of whether authentication will need to be highly reliable, unacceptable levels of false positives/negatives, and whether the reliability will withstand the passage of time. The paper offers a certain degree of flexibility through the various biometric modalities used for identity recognition and transaction confirmation, but does not describe the technology's ability to mitigate biometric drift or inconsistency. The use of tokenization provides a degree of protection by abstracting the biometric data in its raw format. Also, privacy is a significant issue for the technology. The system appears to be observing and interpreting user behavioral data continuously, but provides little elaboration on the user's ability to govern their data, or on the user's right to define their own biometric identity. Finally, the provision of AI assistants and prediction of interests provide additional elements of personalization, but extend the surveillance nature of the proposed solution. This solution raises broader questions about how ethical data about potentially vulnerable users is used, along with the potential for bias in the algorithm. Revanth et al. [10] give a focused study on examining touch-based and other biometric forms of authentication. The study indicates that traditional security methods, such as a PIN or graphical type pattern, endure the need for the user to minimize shoulder-surfing and to combat cognitive fatigue in memorizing, while biometric modalities may be seen as to provide an immediate primary use that gives less vulnerability, but may introduce privacy concerns for users and be a hindrance in practical implementation by high costs and infrastructure requirements. The study also identifies that the use of touch dynamics biometrics is on the rise, together with the exploration of non-touch modalities, specifically if touch dynamics forms are based only on the specification of a user's unique touch interactions—as well as with regard to touch-free modalities, such as facial recognition, iris scans, and even partially contactless forms like voice. The study highlights the emerging work on multi-factor authentication (MFA) access via usable security in stating that MFA access will likely have multiple biometric and behavioral components. The increase in exploration of new biometric systems, which includes neurophysiological signals such as EEG data, has started to get a lot of attention due to their unique nature and the fact that it is hard to forge. EEG-based authentication relies on the dissimilar nature of brainwave patterns of each individual, especially when performing mental tasks. Various studies have proposed methods that utilize features such as past impact values, EEG channel spectral power, and signal complexity measures such as approximate entropy. These features capture both the location-based and time-varying individuality of the brain activity. One of the new methods utilizes a two-phase model of authentication that employs these EEG features to reduce false acceptances and rejections. This method has demonstrated high accuracy in small subject group trials, with zero errors under laboratory conditions when utilizing specific mental tasks. Besides this, applying modified cross-checks helped improve the reliability. Though with encouraging results, current research is constrained by small populations, and additional work using larger populations is required to determine the scalability and long-term integrity of EEG as a biometric modality [11]. We rely more on secure access systems, as traditional methods like passwords, PINs, and keycards are easy to steal, lose, or even overhear. While biometric options like fingerprint and retina scanning are more secure, they are not accessible for everyone, especially people with physical disability. As a result, this led researchers to start exploring brainwave-based authentication as a safer and more convenient option. EEG signals, which reflect the user's cognitive and neural processes, are distinctive and difficult to imitate. Unlike physical typed data, brain signals are not easy to forge with. Systems have been suggested that allow the user to develop personalized brainwave patterns such as combinations of eye blinks, attention levels, and certain brain rhythms (Alpha, Beta, Theta, Delta) as part of the verification process. Some architectures take this further by adding a second level of verification, such as a passkey, following successful EEG pattern recognition. These multi-level systems hold tremendous potential in security and usability. These breakthroughs show that technology connecting the brain to computers might be used in the future for safe and easy login systems [12]. Rahul et al. [13] show behavioral biometrics have been a promising area for improving user authentication for digital systems. Unlike fixed passwords, behaviors like typing style, how someone moves through a system, and gestures offer a live and changing way to confirm identity. Modern systems utilize real-time capture of user device data to examine these subtle patterns through high-end processing modules and ML methods. Recent advancements combine behavioral analysis with adaptive algorithms learning from user behavior over time, not only detecting but also reacting to changes in behavior. By combining real-time detection with smart processing, the system is enabled implicitly and can automatically verify a user without them having to do anything noticeable. This not only makes things more secure but also more convenient and user friendly. It creates a unique biometric profile for each person and keeps authenticating their identity in the background, which helps in preventing unauthorized access. This technique matches widely used practices in smooth user-friendly security, where user experience is optimized without reducing protection. The integration of AI and behavioral analytics thus carries significant promise for future-proof secure access technologies. Biometric recognition systems have attracted major interest in authentication security because they use physiological and behavioral characteristics to determine identity. However, many previous systems suffer from higher false rejection rates (FRRs) to variations in biometrics stemming from environmental, user behavioral, or temporal conditions. This study presents an innovative adaptive framework that avoids limitations of previous systems by utilizing a match determination mechanism, which allows active updates to the biometric template repository. This system allows it to evolve over time based on additional authentication inputs, while providing a pathway for less strict thresholds with an establishing condition that includes authentication-bound time updates, quality, liveness detection measures, and, successfully, multimodal credential checks. Allowing passive enrollment and condition-based updates adapts biometric templates, therefore giving systems a better likelihood to resist variability when it is not detrimental to security. This approach reinforces an essential advancement to biometric systems, creating a pathway to bridge accuracy and adaptability [14, 15]. With the growing requirement for advanced security in secure environments such as bank vaults, conventional methods of authentication are insufficient to resist coercion or deception. Biometric systems have emerged as the favorite since they can provide increased security based on physiological features. Among these, brainwave-based authentication is the best since it is exclusive and cannot be imitated. EEG signals, which are characteristic of a person's brain activity, can be used as a secure measure, particularly in situations where conventional biometrics can be tampered with. Recent research has investigated the possibility of EEG-based systems for authenticating users on the basis of cognitive task responses such as selective and sustained attention, memory recall, and response to varying stimuli. These methods use frequency domain analysis to extract characteristic neural features. Other factors like how long authentication takes and how comfortable it is for the user are also being studied. Research has indicated that such systems can secure vaults most of the time and have a high accuracy rate, especially when tasks based on long-term memory are utilized. Such findings are in favor of the usability of EEG as a secure biometric for high-security applications.

8.2 EEG-BASED AUTHENTICATION SYSTEM

Arias-Cabarcos et al. [16] represent a study of the opportunities proposed by using EEG brainwave patterns and measuring brain activity for authenticating users. This work aims to circumvent some of the limitations associated with classical biometrics like fingerprints and facial recognition, which have issues related to visibility, covert stealing, and irreversible disclosure. EEG does not share these limitations and provides privacy, revocability, and also provides liveness since EEG brain activity is always internal and not visible. The study involved the use of a commercial-grade EEG headset, the Emotiv EPOC+, and tested five event-related potential (ERP) tasks, two P300 based (image recognition) and three novel N400 based (semantic processing) tasks (recognizing a celebrity face, determining sentence congruency, and matching a word pair). Data from 56 subjects was captured in a controlled lab atmosphere and using known methods of EEG cleaning, processing, and operationalization. Power spectral features were extracted across Delta to Gamma bands, and during the work, classifiers were built using three different methods—SVM, Random Forest, and Naive Bayes—with SVM providing the greatest repeatability across all test conditions. The N400:Faces task had the lowest EER at 14.5%, demonstrating repeatability and user specificity that outperformed the attention-based P300 task. The study explored also the continuous trajectory of EER and user experience where participants preferred the more creative, semantic-based tasks. Overall, while the results indicate that low-cost EEG systems can gather EEG data to authenticate users, the need to reach the standard of secure authentication is difficult with current EER levels.

The research in [17] discusses the potential for non-invasive BCI technology that utilizes EEG systems, where the presence, absence, or frequency of certain brain waves in the beta, alpha, and theta frequency bands was found to detect and differentiate both cognitive and emotional mental states. The researchers used the Emotiv EPOC headset and Brain Activity Map software to log the EEG signals that provided real-time responses as subjects performed various cognitive and emotional tasks. Beta waves were shown to have the most predominance in the problem-solving task, where subjects were carrying out what was considered logical mental processing with energy and an increase in cognitive effort, whereas theta waves showed a predominance among subjects during the problem-solving when they were in disorganized thinking or free associate thought patterns and low-focus mental states of transitional cognition. When subjects were in a relaxing state as provided by calming music mixed with nature sounds, alpha and theta waves emerged, which suggested that subjects displayed characteristics of reaching a deep state of relaxation while also gaining low-stage meditative awareness. The research showed that there was no meaningful brain wave activity among control groups that provided a baseline of inactive conditions. Findings suggest that EEG frequency detection could differentiate various cognitive states and provide an available and non-invasive method for monitoring real-time mental states throughout other low-cost EEG systems. This research supports a growing body of evidence that EEG technology can be a means of delivering low-cost solutions for neuro-rehabilitation applications, emotional tracking analysis, and brainwave-based authentication and patterns of control for adaptive human–computer interfaces.

Fladby's [18] thesis investigates the ultra-modern concept—your brainwaves could potentially be used as a password. So, instead of typing something, you would just think of it instead like color, song, shape, or even a word, and the system would identify you based on how your brain reacts. He used a basic and inexpensive headset with one sensor in the forehead region (the Fp1 spot) to measure brain signals from 12 participants who each participated in eight different mental tasks that included relaxing, counting, and imagining. Each participant completed these tasks during three different sessions. The brain signals were recorded and then classified and further examined, both from a shape over time perspective and a frequency point of view (such as Delta, Theta, and Alpha). From each brain signal, 25 features were extracted, such as the number of times the signal crossed zero, or how strong the signal was in different bands of frequency. To assess if this could even work as a possible authentication mechanism, he used two more formal approaches—one where a comparison considered the shape of the brain signal in total (Dynamic Time Warping), the other approach to compare mostly key features.

The paper discusses brain signals captured by EEG (Electrical activity via Electroencephalography) for the purpose of authentication. While methods such as passwords and fingerprints can be effective, each has some aspect of security that is ultimately flawed. Passwords can be guessed, and fingerprints can be spoofed. On the other hand, EEG is difficult to fake, it contains a signal that is unique to each person, and during authentication, EEG reflects the person's unique mental state in real time. The paper discusses the difference between EEG equipment; many use high-end medical devices and others use consumer devices like Emotiv and NeuroSky. The rest of the paper discusses three basic types of authentication protocols: task-based (the user completes mental tasks), resting state (the user is not completing a task, we are just datasetting passive signals), and stimulus-based (the user responds to audio/visual triggers). Turning EEG with brain signals into a user authentication is based on extracting features from EEG time series using a variety of methods from simple statistics to wavelet transformations, and then applying a classification method based on the extracted data to establish examples of training, using methods like SVM, Random Forest, or also deep learning models like CNN and Recurrent Neural Networks (RNNs). There are some systems where people have used motor imagery with CNN where system accuracy would be greater than 95%. Nevertheless, the paper also emphasizes practical obstacles: signals from the brain are inconsistent across moods, levels of fatigue, and stress; brain signals can also be noisy when eyes or muscles shift (to read a message, for example); plus, most people likely would not wear EEG headsets on a customary basis. Table 8.1 gives a comparative study of three modes of authentication, passwords, fingerprints, and EEG-based systems, in terms of selected features. With regard to spoof resistance, passwords are the least resistant to spoofing attacks because they are easily stolen or guessed. Fingerprints do not provide much resistance to spoofing. EEG-based authentication systems are highly resistant to spoofing attacks because brain signals are very complicated. As it is unique for every individual, passwords depend on user-selected passwords and provide

TABLE 8.1 Comparison of traditional vs EEG-based authentication methods

FEATURE	*PASSWORDS*	*FINGERPRINTS*	*EEG-BASED*
Spoof resistance	Low	Medium	High
Uniqueness	Medium	High	Very high
Real-time verification	No	No	Yes
User convenience	High	Medium	Low
Hardware requirement	None	Fingerprint scanner	EEG headset
Vulnerability to theft	High	Medium	Low

little bit of uniqueness to the user. Fingerprints provide high uniqueness to a person because fingerprints are unique to each individual. Signals derived from an EEG provide very high uniqueness because an individual's EEG signal represents the individual's neural activity. All forms of authentication do not provide a real-time verification option (passwords and fingerprints require additional user input), whereas the EEG authentication system can be used to verify the user based on brainwave data provided in real time using a live EEG signal. In terms of ease of use, passwords are the easiest to use, followed by fingerprints, and EEG systems are the least user-friendly authentication system since they require equipment. In terms of hardware requirements, passwords do not require hardware, fingerprint scanners require a fingerprint scanner, and EEG authentication requires an EEG headset to be worn by the user. In terms of theft, fingerprints provide medium theft capability, and EEG authentication has low theft capability and provides enhanced security against identity theft.

Also Table 8.2 gives a comparative summary of three EEG-based authentication protocols, including Task Based, Resting State, and Stimulus Based, along with their description, expected use cases, strengths, and weaknesses. Task-based protocols allow the user to think through a range of predefined mental challenges, such as completing math problems or thinking about particular actions. Because task-based protocols yield a user-specific and a recorded or clearly defined way of stimulus, they are most useful in mental challenge-based systems, especially in high-security applications producing relatively high uniqueness based on work assigned to particular tasks. However, they require significant intellectual effort and would not be appropriate for frequent or day-to-day authentication needs. Resting state protocols record the brain activity while the user is at rest and typically idle or relaxed, essentially in a "non-doing" state. This type of protocol is particularly desirable as it is passive and can potentially be entirely continuous for the user. That said, it assumes users put in less effort and have minimal involvement, but directly reflect on varied brain conditions with expected lower accuracy for user-independent detection and making the user's personal traits even harder to tell apart from others. Stimulus-based procedures operate through the user's response to external stimuli such as visual or auditory stimuli, and can constitute graphic-based actionable items applied in challenge-response procedures; they similarly utilize stimulus patterns that engage and reflect user responses in contrast to non-response-based task stimulation problems used in the task-based protocols. While stimulus-based protocols increase discriminative power to the protocol (due to external stimulus-induced brain characteristics), complexity can increase due to poorer control or unintended near-adverse distracting environments from the user. Each address has different compromises between accuracy and usability. Then, of course, there is the encumbering issue of data privacy, as brain data could expose the personal mental states of users. As a forward-looking suggestion, the paper advances myriad combinations of EEG and additional biometrics (e.g., face, voice), constructing EEG devices that are more ergonomically

TABLE 8.2 Types of EEG-based authentication protocols

PROTOCOL TYPE	*DESCRIPTION*	*TYPICAL USE CASE*	*STRENGTHS*	*WEAKNESSES*
Task-based	User performs specific mental tasks (e.g., math, motor imagery)	Cognitive challenge authentication, high-security systems	High uniqueness from task-specific brain signals	High user effort; may be impractical for frequent use
Resting state	EEG is recorded while the user is idle or relaxed	Continuous passive authentication	Low user burden; minimal interaction	Lower accuracy; harder to differentiate user's reliably
Stimulus-based	User responds to visual or auditory stimuli	Media-based, challenge-response scenarios	Stimulus-linked brain signals increase discriminative power	

acceptable, smarter, and explicable; AI models; and documenting standardized datasets which would assist the shift to practical use from laboratories.

Table 8.3 gives a summary of the various feature extraction and classification methods that have been utilized for EEG-based authentication, by comparing them based on their developing signal domain, computational cost, and the classifiers they were usually used with. Time-domain statistics, such as mean and entropy, are developed in the time domain, have a low computation cost, and are typically executed with classifiers such as SVM and decision trees. PSD, usually developed in the frequency domain, has a medium computational cost and is primarily used with the k-Nearest Neighbors (kNNs) or SVM. Wavelet transforms like USA DWT, CWT analyze time and frequency components simultaneously, at a medium computational cost, and are usually developed with ensemble or Random Forest classifiers. Autoencoders have a high average development complexity and power, and are routinely developed with a CNN, LSTM or Deep Belief Network for classification. FFT works in the frequency domain and is usually a low to medium computation cost, and is often used with classifiers such as SVM or Naive Bayes. Common Spatial Patterns (CSP) are used for spatial filtering, have a medium to high computational cost, and are often built with classifiers such as linear discriminant analysis (LDA) or SVM. Overall, each method presents various compromises when considering computational costs, any potential signal characteristics, and classifier compatibility. In general, the paper identified great promise for EEG-based authentication but argued several technological and ethical barriers remain on the front stage of EEG-based authentication [19].

The survey research [20] reviewed several deep learning models that could be put to use regarding EEG data analytics, particularly focusing on CNN, RNN, and hybrid models. CNN is arguably the best network architecture to extract features from raw EEG signals, or even an image of a frequency spectrogram. CNNs can use convolutional filters to help characterize and pull a spatial hierarchy of features. This indicates that CNN's capabilities were useful in classifying brain activity and detecting variability of anomalies. RNNs have been shown to provide value when working with sequential data and are used in predicting an event that has a temporal component to it, like a seizure. One of the ways RNNs are able to predict so successfully is manipulating a hidden state to maintain information about past inputs. Figure 8.1 gives an example of an end-to-end EEG classification pipeline using a hybrid

TABLE 8.3 Feature extraction and classification methods in EEG authentication

FEATURE EXTRACTION METHOD	*SIGNAL DOMAIN*	*COMPLEXITY*	*COMMON PAIRING CLASSIFIER*
Time-domain statistics (e.g., mean, variance, entropy)	Time	Low	SVM, logistic regression, decision tree
Power spectral density (PSD)	Frequency	Medium	k-nearest neighbors (kNN), SVM
Wavelet transform (e.g., DWT, CWT)	Time-frequency	Medium	Random Forest, ensemble methods
Autoencoders (deep feature learning)	Learned feature space	High	CNN, LSTM, deep belief networks
Fast Fourier transform (FFT)	Frequency	Low to medium	SVM, Naive Bayes
Common spatial patterns (CSP)	Spatial filtering	Medium to high	Linear discriminant analysis (LDA), SVM

FIGURE 8.1 End-to-end EEG classification pipeline using a hybrid CNN-LSTM architecture.

CNN-LSTM architecture for classification or abnormality detection. The input information first undergoes initial data acquisition as EEG signals. After that, the acquired EEG signals require preprocessing for cleaning and standardizing the signal.

Following the preprocessing step, the processed EEG signals are run through multiple CNNs, which intelligently extract salient spatial features from the EEG signals. The salient spatial features extracted from the CNN are then passed on to LSTM networks, which model the temporal dependencies and sequence information within the EEG signals. Finally, the processed representation created from the CNN and LSTMs is passed into the output layer where the system will classify or detect abnormality. The combination of CNN and LSTM networks is useful because they allow the ultimate system to understand both the spatial and temporal characteristics of EEG signals, assisting in smart, robust, and repeatable decision-making. In addition, LSTM RNNs are formed in a way that the memory neuron will not forget important input information, and will be presented with new input to consider. LSTMs help RNNs by providing gated features that permit or restrict items in the memory neuron. Hybrid models provided the best of both worlds using the features of CNNs to explore spatial elements of the data and the temporal aspects using LSTMs, as both NNs have strengths in different NN space. Also Figure 8.2 provides a diagram of architectural designs for processing EEG signals between CNN, RNN, and the hybrid CNN-RNN mode.

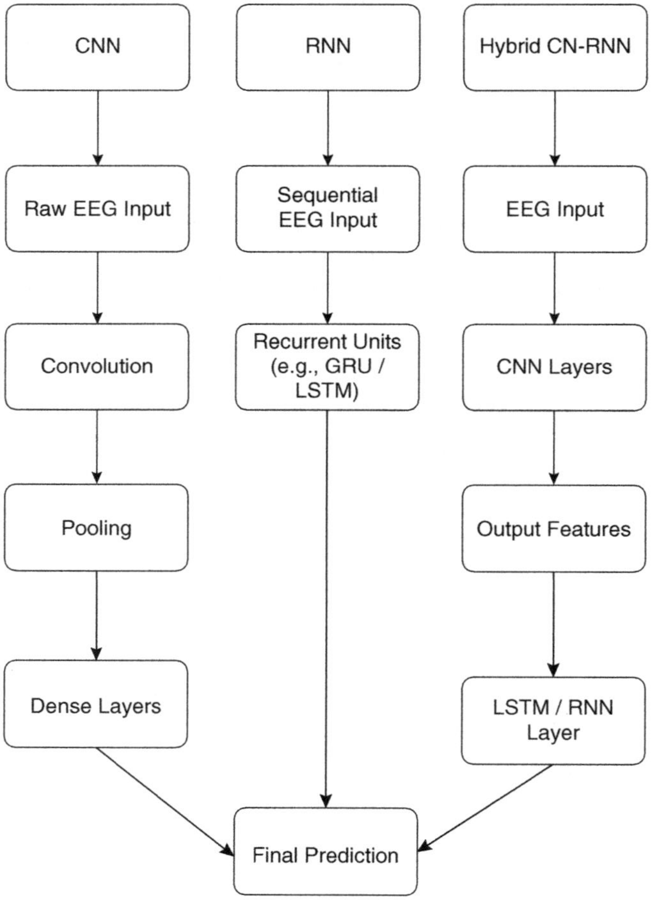

FIGURE 8.2 Architecture comparison between CNN, RNN, and hybrid CNN-RNN models.

Whatever the model used (pure CNN or pure RNN), the architecture of the model needs to include additional dense and functional layers leading up to the predicted output. The CNN-based model relies on raw EEG input that is processed through convolution layers designed to extract spatial features of the EEG input. The model uses a few steps to pick out useful information and combine it so it can give the final answer. The RNN model, on the other hand, processes sequential EEG input through recurrent units (GRU or LSTM) designed to capture temporal dependencies, leading directly to the built-in output. With respect to the hybrid CNN-RNN model, the model capitalizes on the strengths of both neural architectures processing spatial features of the EEG input through CNN layers and then feeding those features into RNN layers that model the temporal patterns in the EEG. The process of spatial and temporal learning through the hybrid CNN-RNN is appropriate for the higher-level structure and complexity of EEG classification tasks. However, potentially using both networks will help to increase accuracy in prediction about brain activity or potential prediction of anomalies from other non-deep learning models, which may improve identifying anomalies through diagnosis and improving clinical outcomes for BCI applications.

The study [21] examines the possibility of utilizing EEG signals as a biometric measure for user identification and verification, given the uniqueness and difficulty of replicating EEG measurements. The approach applied for EEG user identification and verification utilized noise cancellation following ensemble averaging and low-pass filtering, followed by feature extraction using a wavelet packet decomposition (WPD) and a classification approach following an ANN; however, an alternative machine neural approach would improve architecture. The system architecture implemented consists of four basic components: (i) Signal acquisition, (ii) preprocessing area where signal identification and anxiety/incident are measured, (iii) feature extraction-signal to text via WPD, and (iv) classification applied based on added value. Their study explored four application scenarios with 32 adult participants (11 females, ages 18–25) and was able to identify consistent EEG signals from six midline electrode sites such as Fpz, Cz, Pz, O1, I2, and O. In Scenario I, the focus is on identifying all subjects, and the accuracy was noted as between 5.75% and 10.68% correct recognition. Scenario II uses a side-by-side identification system. The subject-estimated accuracy improved between 28.71% and 36.27%. In Scenario III, the subject is successful, with subject accuracy of 94.04% correct recognition. Finally, in Scenario IV, a small group or authorized personnel occurrence with an estimated 70.06%–99.20% correct recognition was identified. Table 8.4 gives details on the EEG-based user identification system, describing its operational components such as signal acquisition, preprocessing, feature extraction, and classification. During the signal acquisition phase, the EEG data is acquired from six cortical electrode sites on the head—Fpz, Cz, Pz, O1, I2, and O—to collect information from different brain areas. In the preprocessing stage, a range of methods is applied to the raw EEG data to enhance its potential quality, including ensemble averaging and low-pass filtering, to remove noise elements in the signal and the detection of incident or anxiety-induced distress, which may have impacted the signal precision. Next, during the Feature Extraction stage, the amplitude features of the EEG data will be analyzed using WPD, extracting relevant features and converting the complex time-domain EEG signals into vector-like text data

TABLE 8.4 EEG-based user identification system architecture

COMPONENT	*DESCRIPTION*
Signal acquisition	EEG signals collected from six midline electrode sites: Fpz, Cz, Pz, O1, I2, and O.
Preprocessing	Noise cancellation using ensemble averaging and low-pass filtering. Also includes detection of anxiety/incident events.
Feature extraction	Wavelet Packet Decomposition (WPD) is used to extract relevant features from EEG signals and convert them to text-like representations.
Classification	Classification is done using Artificial Neural Network (ANN). The study suggests trying alternative neural models for improved performance.

TABLE 8.5 Scenario-wise EEG user identification accuracy

SCENARIO	*DESCRIPTION*	*ACCURACY (%)*
Scenario I	Identification of all 32 subjects using EEG data from six midline electrodes (Fpz, Cz, Pz, O1, I2, O).	5.75–10.68
Scenario II	Side-by-side subject identification with moderate improvement in recognition accuracy.	28.71–36.27
Scenario III	Subject-wise recognition from EEG signals with a high degree of accuracy.	94.04
Scenario IV	Small group or authorized personnel identification with strong recognition accuracy.	70.06–99.20

that is ML-compatible. Finally, during the classification stage of the process, an ANN will be used to identify users based on the extracted features. This research paper has also noted the potential benefits of other NN models if the user identification system performance or accuracy needs to be improved. Table 8.5 gives a brief overview of how accurately an EEG-based user identification system can perform in four different scenarios as mentioned in-depth previously.

Shams et al. [22] describe how our brainwaves, registered as EEG signals, can be taken as a secure, one-of-a-kind representation of our identity. Brainwaves are incredibly hard to copy and therefore make for a promising new biometric form of authentication. The authors summarize a wide, directly related body of literature that has used ML functions to classify brainwave patterns—varying in response to whether a person is physically moving, imagining moving, resting, or responding to the presentation of external stimuli such as pictures. They summarize a lot of the results and find that deep learning was typically used and typically led to the best results, though there are still a number of technical challenges involved from signal change over time to the basic technical use of EEG, and little understanding or control over the technical complexities involved. The paper concludes with suggestions for continued research to improve the accuracy, efficiency, and usability of the technology. This research presents two novel classifiers: Isolation Forest (IF) and Local Outlier Factor (LOF), which to date represent the first time they have been used for user authentication based on EEG signals. The paper states that model parameters are important for the security of the system as a whole. They achieved this at almost 100% precision and just above 75% recall, calling them usable models. The user authentication occurs in two phases: the enrollment phase when data is collected from the user and then used for training the model; and the authentication phase when model testing uses new measurements. Multi-Class classifiers, such as SVM, Random Forest (RF), and kNNs, instantiate both positive and negative (impostor) samples in the user authentication task, but one-class classifiers, including OC-SVM, are able to instantiate a user with just the positive samples, making them advantageous for tasks without impostor samples. The authors developed a hybrid system, which combines one-class classifiers and multi-class classifiers, achieving an accuracy of 82.3%, precision of 91.1%, recall of 75.3%; additionally, they identified an exam period that included P8, T8, FC6, and F4 for the users, and particular frequencies of the user's high-frequency brainwaves like gamma and beta for the authentication. Table 8.6 compares the classifier types used with EEG-based authentication by describing the classifiers and what they try to accomplish. One-Class classifiers such as IF, LOF, and OC-SVM identify users using positive samples only. The advantage of one-class classifiers is that they can be used to identify users without the impostor data, because they learn to detect class based on the known user patterns instead of based on impostor patterns too. Multi-Class classifiers, such as SVM, RF, and kNNs, utilize the impostor data in conjunction with the positive samples to separate users and attackers, and have better discrimination. A Hybrid classifier method takes advantage of both one-class and multi-class models. The first stage identifies user characteristics using one-class methods, and then improves discrimination through multi-class classifiers. The two-stage process increases accuracy and capabilities of the overall system and works

TABLE 8.6 Classifiers used and their purpose in EEG-based authentication

CLASSIFIER TYPE	*CLASSIFIERS USED*	*PURPOSE*
One-class	Isolation Forest (IF), Local Outlier Factor (LOF), OC-SVM	Identify users using only positive (genuine) samples. Useful when impostor data is unavailable.
Multi-class	SVM, Random Forest (RF), K-nearest neighbors (kNN)	Use both genuine and impostor samples to differentiate between user and impostor.
Hybrid (proposed)	One-class + Multi-Class	Combines strengths of both classifier types for improved performance in EEG-based authentication.

well with real-world applications in EEG-based authentication. Next studies will explore integration with other types of biometrics, develop deep learning approaches to develop a more robust user authentication system, and test an impostor attack on the EEG user authentication system using Generative Adversarial Networks (GANs) [23].

Tot et al. [24] introduce a hands-on software platform called elEEG that teaches people how brainwaves could be used to provide a secure form of user authentication. elEEG comprises two main components. One is a low-cost software tool for collecting brainwave data using the NeuroSky MindWave 2 EEG headset, and the other is a web tool for visualizing and analyzing the data. Participants engaged in four short tasks—relaxing, reading, solving math problems, and viewing emotional image inputs—while their brain signals were recorded across five types of brainwaves: alpha, beta, gamma, delta, and theta. To establish the elEEG's efficacy as a platform for learning, the researchers chose two cohorts: one was taught using traditional lecture techniques, while the other cohort used the elEEG system. The elEEG cohort had higher performances in practice-based tasks, but they also reported feeling more motivated, engaged, and had a greater understanding base of how brainwave data works. The research shows that EEG technology can be used to meet security requirements for highly secure, free-flowing, and completely personalized user authentication systems, and the use of elEEG can enhance the access and engagement of learners to such complex and comprehensive biometric concepts.

The paper [25] describes a real-world experiment that shows how brain signals can serve as secure, password-free authentication through a use case study. As part of a research study, 49 participants comfortably lived with and wore EEG headbands for 7 days. During each session, the participants watched a brief stream of visual images that were designed to elicit unique and individualized brain responses from participants. The brain responses were compiled into "brain IDs", which are unique user biometric signatures based on their neural activity. Brain ID provided a reliable method of authentication without any setup and commercial-level performance without the need to be an expert or have a lab environment. The article presented the benefits of brain-based authentication: first of all, brain signals are "invisible" biometrics that are non-invasive, hidden, highly complex, and indistinguishable; therefore, brain-based signals are practically more difficult to replicate or rather steal from, compared to other more traditional biometric methods that are fingerprints or facial recognition. The study also shows it is feasible to integrate EEG sensors into modern wearables that are currently or almost being used, such as earbuds, VR headsets, AR glasses, which could lead to fully isolated systems that could potentially streamline continuous authentication without users needing to do anything extra on their part but simply during everyday interactions.

The research in [26] investigates the means through which ML approaches can be used to exploit spatial patterns in brain wave data corresponding to episodic memory. The authors focused on a region of the hippocampus called the CA1 in rats, which recorded Multiple-Unit Activities (MUA; the firing of multiple cells) during four different experiences: social stress in the presence of a male or female rat, restraint stress, and an encounter with a novel object. The authors examined these MUA records in whole, using various classifiers of deep learning including SVMs, CNNs, VGG16, and hybrids. Previous research has established the role of hippocampal ripple waves in the processes of memory consolidation, and the authors provided evidence that they might be able to use this information for their study based on utilizing the Short-Time Fourier Transform (STFT) to preprocess the data. This confirmed all the appropriate low (delta) and high (beta) waveforms recorded. Deeper networks like VGG16 in combination with SVM provided the highest classification accuracies of just over 95.6% for one episodic experience unto another. In addition, the authors used Grad-CAM to visualize those portions of the MUA signal that contributed most to the respective classification, typically consisting of ripple-like waves, that were specifically related to respective episodic memories. This study has advanced what we have learned, because it demonstrates that deep learning can be used to recognize episodic memory-related neural patterns, derived from complex neural representations with MUA data, whereas previous EEG studies have focused more broadly on cognitive states or general neural rhythm network coordinations of brain function. This study may also have a niche in the development of tests for diagnosis in memory disorders to subsequent work such as BCIs that includes episodic recollection. It should also be noted that there were limitations to this work in terms of small sample size and species specificity, which suggests a need for broader validation.

The EEG-based BrainPrint system uses brain signals to unlock smartphones, rather than recognizing typed passwords, providing a truly hands-free and secure authentication method. Users wear a Unicorn EEG headset that reads their brain activity (P300 signals) as they concentrate on blinking numbers. The P300 signals are detected, processed, and sent to the smartphone by Python and AirDroid. While developing the interface, users watch short video recordings while their brain signals are captured, then processed in MATLAB, to create a unique BrainPrint biometric pattern instead of a password. The BrainPrint pattern is then provided, or added to the password, to authenticate the user using a ML model (kNN). The BrainPrint system requires two PCs that fit together, at least one smartphone, a headset, and the ability to read brain signals in a noise-free environment for no distractions from nonsense activity. This BrainPrint system offers people with disabilities the option to use brain signals; even more importantly, it offers to enhance privacy because there is no physical input [27].

The paper [28] describes a clever and highly secure method of authentication: utilizing the images paired with your unique brain waves to authenticate your identity. When you're attempting to login, the test shows you several images on a screen, where each image flashes or flickers at its unique speed. One of the images is a "password" picture that is associated with your identity. When you're fixated on that image, your brain will emanate distinct rhythms and patterns of brain activity that can be monitored with a brain wave sensor like an EEG device. As a user, the system will compare your brain waves with the same brain waves it recorded previously, when you first set up your password. If the rhythms and patterns belonging to the correct image-password match your brain waves, it is identifying you as the person, if not, access is denied. For other security, we can vary the images in the test, so that every time you log in, the images will be different, thus making it even harder for a user to imitate or impersonate you. By leveraging the uniqueness of your brain activity along with images, this method provides a highly secure, personalized, and strong authentication for your devices.

The invention in [29] creates an intelligent, secure means for confirming a user's identity before authorizing access to an online service. When the user attempts to log in, or perform a transaction, the system will generate the unique code, together with one additional piece of information, a clue associated with the user's clue to a secret pair, often a custom riddle or other clue. This unique code and clue, accompanied by a security token—a message authentication code or "MAC", to provide some protection against tampering—will be sent to the user's device. The user must confirm the unique code before he/she can view the clue, validating that it is indeed the expected clue previously associated with the unique code the user confirmed. The user then unlocks the device using a fingerprint or face scan, followed by correctly entering the answer to the clue. If all components check out, access is granted. The unique login codes or transaction codes are also unique to every login transaction, which will also help to provide a unique level of personal security for the individual user.

The paper [30] demonstrates the issues with conventional biometrics, which can have relatively high-security risks due to the ability for replication, such as fingerprints and face recognition. It also shows that EEG potentially provides similarly secure biometric alternatives because they are unique to individuals while they are alive and neither could be reproduced in a deceased state. The biggest barrier to success with EEG is the inherent variability in EEG data, based on their mental states that make continuous authentication impossible to consistently achieve. To overcome any concerns with variability, authors show the use of deep breath exercises as a way to stabilize their brain state and facilitate stronger alpha wave patterns. The authors used EEG data from 15 subjects with structured levels of deep breathing and operationalized deep breathing with preprocessing of the EEG signal such as filtering and Independent Component Analysis (ICA), and focused on alpha and beta wave feature extraction. The study found high classification accuracy counts of 92% using SVM and 90% using NN in the breath-holding phase, and also claims deep breathing provides stronger unique identifiers, and suggests deep breathing as a method in next-generation biometric EEG authentication systems across many mental states.

8.3 EEG EMOTION-BASED AUTHENTICATION

Emotion detection is a rapidly evolving field, transforming from traditional psychological approaches to current ML-based systems. Traditional methods based on rules provided starting points for development yet do not have the flexibility and scalability of ML-based methods. Supervised methods including SVM, Decision Tree, and NNs have demonstrated feasible results with desired outputs based on data describing emotions from facial, vocal, and textual inputs. Unsupervised methods such as clustering, Dimensionality Reduction methods maybe deployed. CA and t-SNE complement the supervised collective feature extraction and inference based on pattern identification. Deep learning algorithms, especially CNN and RNNs, have facilitated a new era of automated feature learning and utilization of sequential data. The literature recently emphasizes the growing importance of multimodal systems that simultaneously utilize visual, auditory, and textual information to provide information regarding a person's emotions in a more robust and reliable way. Although emotion recognition systems can span applications ranging from mental health, marketing, human–computer interaction, and surveillance, technical and ethical challenges remain. First, representation distance, whether created due to variance like cultural distance or contextual distance: the effectiveness of the emotion detection system is also significant. Addressing issues of data privacy and as potential outputs are generated to sensitize the far more complex connections to a person and their dis-positioning through the use of postulated substantial identity and emotional representations will present significant challenges for further research in paradigm change. In summary, the paper [31] has discussed the timeline for advancements in ML to detect emotions, which will further broaden current and future emotion detection systems. Following the paradigm shift informed by due diligence from any emergent ethical ramifications across the spectrum of data representation, interpretation acknowledges how ML is applied across interactions with numerous roles; this is a challenging road to confront in our discipline and to research mappings.

The paper [32] proposes a novel, practical way of personal identification using brain signals—by analyzing a person's blink. Rather than requiring complex cognitive tasks that could be influenced by emotional or physical states, the researchers utilized the brain signals created by the consistent and reflexive act of eye blink when study subjects blink. Table 8.7 shows a diverse selection of variability on biometric modalities, which is all based on input conditions, accuracy, the sensitivity toward emotional

TABLE 8.7 Comparison between modalities

BIOMETRIC TYPE	*INPUT REQUIRED*	*ACCURACY*	*AFFECTED BY EMOTIONS/ PHYSICAL STATE*	*COST/ COMPLEXITY*
Fingerprint	Finger placement	High	No	Low
Face recognition	Camera image	Medium	Yes	Medium
EEG (cognitive tasks)	Mental activity	Varies	Yes	High
EEG (blink-based)	Eye blink signal	97%	No	Low

TABLE 8.8 Blink signal mapping and features

FEATURE EXTRACTED	*DESCRIPTION*
Blink ERP	Evoked brain response generated by the reflexive eye blink, captured as a time-locked signal.
Spatial distribution	The variation in blink signal amplitude across different EEG electrode positions on the scalp.
Temporal consistency	Consistent blink waveform patterns across time, indicating repeatability.
Subject-specific signature	Unique features in the blink signal that vary by individual and can be used for authentication.

or physical conditions, and cost/complexity practices. Fingerprint recognition requires an individual to simply place their finger on a reader to identify/reject them with high accuracy on a reader that is not impacted by emotional/physical changes with low cost/complexity. Face recognition is a camera input that has a moderate accuracy that changes due to emotions or physical/health conditions and has a moderate cost/complexity. EEG based on cognitive tasks identifies cognitive mental activity, has variable accuracy, but is highly sensitive to quick emotional or physical changes and requires a high cost and complex system. On the other hand, EEG blink-based authentication uses eye blinks to produce 97% accuracy when it identifies a person while being unaffected by emotional or physical conditions, which is inexpensive and is less complex.

Furthermore, Table 8.8 presents some distinctive features of blink signals from EEG-based biometric devices, which are important for user identification and user authentication. Blink ERP is the evidence of the brain's triggered response from a reflexive eye blink, which records the time-locked signal from the EEG and indicates neural activity. Spatial distribution looks at the differences in amplitudes of blink signals as recorded by the EEG electrodes in different positions on the head; this spatial analysis will provide an understanding of the spatial characteristics of the blink response. Temporal consistency looks at the temporal consistency of blink waveforms; it is important to ensure the signal is dependable and replicable across time and during several trials. Subject-specific signature highlights aspects of the blink signal that allows us to distinguish subjects, which provides more features that will help build a strong blink authentication system that relies on neurological, spatial, temporal, and personal signal characteristics.

Figure 8.3 represents the Blink EEG Authentication Flow, a multi-phase user authentication process based specifically on EEG signals induced by raw EEG blinks. The Blink EEG Authentication Flow is initiated when the EEG signal is captured in the EEG signal acquisition phase. That is when the brain signals due to the blinks will be recorded. The captured, raw EEG section will then go through a preprocessing step to remove noise and artifacts, etc., so that cleaner data undergoes for processing. After processing the raw signal, the EEG signals collected during the user's blinks will be mapped into a similar form. Once the EEG signals are mapped, the feature extraction process will begin. This is where relevant features or characterizations will be extracted from the user's EEG

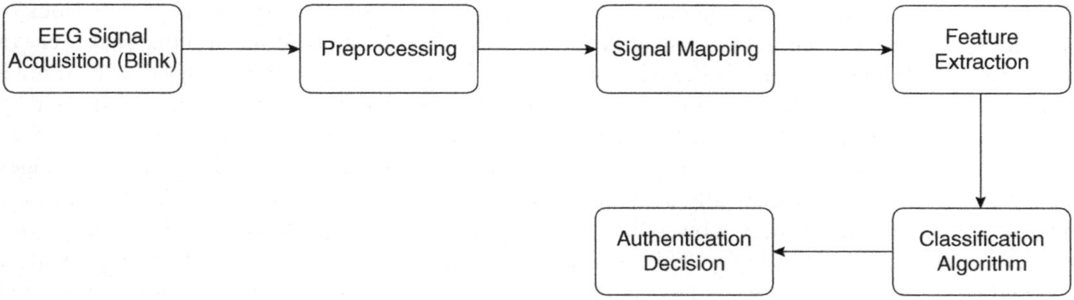

FIGURE 8.3 Blink EEG authentication flow.

biometric recognition and measured against other patterns/profiles from EEG responses. Once feature extraction is completed, these features are then put into a classification algorithm that has been specifically trained with known patterns on users' EEG profiles. Once a classification is made, an authenticated decision can be evaluated. Each phase of the Blink EEG Authentication Flow provides a secure, reliable, and personalized type of biometric authentication based on external brain signal analysis. Overall EEG-based blink is a viable option for both reliability and practicality in a biometric authentication system. With a low-cost EEG headset, the researchers recorded brain signals while subjects blinked, and were ultimately able to create an algorithm that spatially mapped the signals received from individual brain signals captured while blinking, for the purpose of matching individual signals as a form of authentication. Their system exhibited astonishing performance, achieving 97% accuracy, and was unaffected by the physical and emotional elements the users experienced while blinking. This was then determined to be an easy-to-use, effective, and secure option, compared to the more conventional biometrics options.

Emotion recognition is advancing in the fields of computing, healthcare, and human–computer interaction [33]. Apart from traditional methods, which are dependent heavily on facial expression and text analysis, they are less effective depending on personal judgment, which can be easily deceived by signals that are not real. In real-time emotion recognition, EEG appears to be a real alternative modality and provides a more objective means of measuring what is happening in the brain by monitoring brain wave activity. Recently the literature has started to report the use of deep learning models including LSTM and GRUs, which are superior to the traditional ML models by identifying temporal dependencies in EEG data accurately. According to Chandanwala et al. [33], the models can achieve high performance accuracy to 97% to distinguish emotional states. The performance of their models was significantly improved by combining EEG with multimodal data, or taking advantage of its deep network ability. The authors also pointed out that the three key preprocessing methods that employ for feature engineering, Butterworth filtering, PSD, and Spectral Entropy (SE) are proven to be effective. In many papers that claimed to reduce dimensionality before data was analyzed, PCA was used by authors, and while it is an appropriate and reasonable approach it does remove some of the integrity of the data presented and may not be a good way to have a defense for an event. We should not diminish the outcomes that are very promising; however, limitations exist because we are not able to generalize in the real world because of our small activity and context-specific datasets. Chandanwala et al. [33] concluded that future investigations and work should look more at larger datasets and variety in focus in feature engineering; and too much focus on accuracy and insufficient examination will not assist us to understand questions we are seeking.

Omar et al. [34] prove that brainwave frequencies carry valid indicators of internal states, not just for cognitive recovery but also for emotional recognition as well. The paper also tells us about the technical signs such as using RPR (Relative Power Ratio), frequency bands, and clustering methods, all of which are used for EEG-based emotion system design. Researchers recorded EEG signals from 74 individuals recovering from ischemic strokes and grouped them based on how far along they were in their recovery—in the beginning, intermediate, and final stages. By evaluating the brainwaves using a method called RPR, which looks at very specific frequency bands like beta, alpha, theta, and delta, they also noticed clear trends that matched each recovery group. For instance, patients in the final stages showed more beta activity, which is linked to more active thinking, while those in their initial stages tracked more alpha activity, typically associated with a relaxed or rather calm state. Using statistical and cluster evaluations, the study showed that these brainwave patterns could meaningfully distinguish between different levels of recovery. This approach could be a useful tool in monitoring progress over time and designing specific treatment plans for stroke recovery. Table 8.9 analyzes frequencies and RPRs for each band during stroke recovery. The dominant band is noted and the RPR trend is presented with graphs for dominant bands. From the data collected it is determined that the dominant band for the initial stage of recovery is alpha, which is consistent with strong neural activity for a calm or relaxed mental state. The RPR trend is noted to be high for this phase of restoration. At the intermediate stage of recovery, the dominant band is theta, with a transition phase associated with drowsy or transitional mental state and a RPR trend of moderate, which shows a decline in responsiveness of the brain. In the final phase of recovery, the dominant band is beta and suggests an alert or active thinking mental state, and a high RPR indicates an improvement in brain function. Looking at the dominant frequency band progression from one stage of recovery to another shows the engagement and dynamic transitional changes as the patients recovered from their stroke.

Figure 8.4 shows the distributions of EEG frequency bands of Alpha, Theta, and Beta across three stroke recovery phases, Initial, Intermediate, and Final. In the initial recovery phase, movement activity present was believed to produce typically seen activity in the EEG. In this stage, the most dominant

TABLE 8.9 Analysis of dominant EEG frequency bands and RPR trends across stroke

RECOVERY STAGE	*DOMINANT FREQUENCY BAND*	*ASSOCIATED MENTAL STATE*	*RELATIVE POWER RATIO (RPR) TREND*	*KEY OBSERVATION*
Initial	Alpha	Relaxed/calm	High	Reflects early-stage reduced activity
Intermediate	Theta	Drowsy/transitional	Moderate	Indicates cognitive adjustment phase
Final	Beta	Active thinking/ alertness	High	Strong beta indicates recovery progress

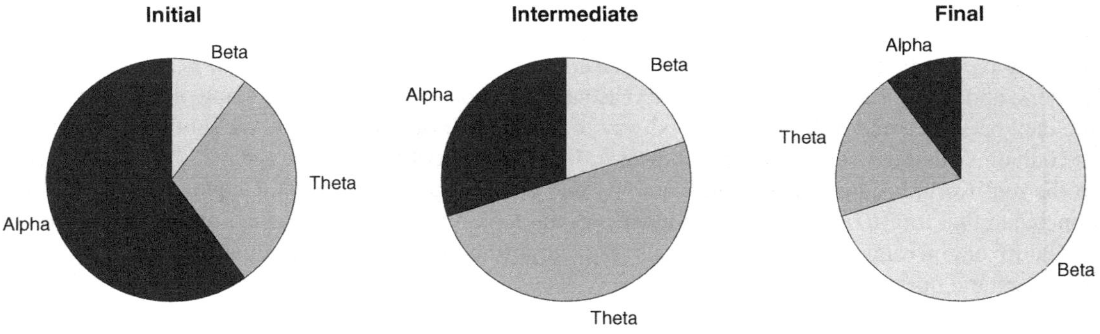

FIGURE 8.4 EEE frequency bands.

band present was Alpha, which suggests a calm relaxed state of mind. The theta band appears, proportionately, with Alpha present. The Beta band was also present, but in smaller proportions. The Intermediate stage shows the most dominant band being theta. When theta is the most dominant band understood the subject is in a state of transition or drowsiness of consciousness. Theta is present with reduced ratios of Alpha and Beta bands and altered ratios. The Final stage EEG shows Beta as the most dominant band. Observed from this phase is a state of alertness and actively thinking. The Alpha band is not dominant and the Theta band is present excessively in very small proportions. This depiction of mental activity collectively depicts the progression of recovery; the stage depicted reflects transition from relaxation and non-active mental activity toward high mental activation in terms of the individualistic progression within stroke recovery.

The study [35] explored the brain and emotional responses to five tastes like sweet, salty, umami, sour, and bitter using EEG data and subjective emotional ratings. A total of 28 healthy young adults participated in the study, during which each participant was individually exposed to taste solutions in a laboratory setting while EEG measurements were recorded. The results revealed that each taste elicited unique brainwave patterns within different frequencies, especially in the beta and gamma frequency bands, because they relate to processing and self-awareness. Sweet taste produced significantly greater gamma frequency activity than umami and bitter taste, while umami and salty taste had stronger beta activity compared to sweet. After all the taste exposures, participants rated sweet taste as the most pleasant and dominant taste, while bitter taste was rated as the least pleasant taste and often confused with sour taste, which may relate back to the deep-rooted nature of experienced aversion for sour and bitter tastes. Overall, the emotional ratings based on valence, that is pleasantness, and dominance, that is sense of control, were significantly correlated with the brainwave activity for every taste except salty. For example, sweet taste led to lower levels of relaxation, which refers to delta or theta wave activity, but greater dominance-based brain activity, and sour taste revealed positive correlations with brain states for relaxed brain and thought activities. Additionally, it was evident that most participants struggled to recognize umami taste, often describing it as unfamiliar or strange. While unable to measure some deeper regions of the brain, and being somewhat limited by a small sample size and manually timed stimuli, this study adds to our understanding of brain processing involving tastes and emotions. It also includes mention of umami, which is commonly overlooked in studies similar to ours, which would seem to provide a fuller picture of taste processing. We hope that this research will enhance future practice in nutrition science, food product development, and sensory marketing, by describing taste in relation to instantaneously measured brain and emotional activity.

The paper [36] is a non-invasive method for enhancing human brain function, including learning speed, memory storage, emotional regulation, and focus. The system is based on EEG to capture the brain's own electrical patterns in real time using sensors on the scalp. The data captured from the sensors can provide information on the user's past and current emotional and cognitive state. However, instead of relying on traditional brain stimulation techniques, this system utilizes adaptive stimulation with sound waves, light flashes, or low-level electrical currents to either emulate or enhance brainwave states usually associated with high-performance or emotionally stable states of mind. What is special about this system is its closed-loop mechanism. Not only does the system read brain activity, it also responsively utilizes ML algorithms to adapt the stimulation based on what the user needs in real time. So, for example, if the user is tired, distracted, or anxious, the system will alter its output to help guide the brain into a more focused or relaxed state. In essence, it is nudging the brain toward its optimum operational state. Table 8.10 presents an adaptive EEG-based neurostimulation system and its core components and tasks in terms of performance enhancement using brain monitoring and stimulation in real time. It starts with acquiring EEG signals with non-invasive scalp sensors that detect brain activity. Next, the data is decoded, through classification, to identify EEG signals that correspond to emotional or mental states. Then, stimuli that define its modalities will adjust brainwave patterns. The adaptive aspect would then involve ML algorithms to adaptively tune the stimulation according to behaviorally

TABLE 8.10 Key components and functions of the adaptive EEG-based neurostimulation system

COMPONENT	*DESCRIPTION*
EEG signal acquisition	Sensors placed on the scalp record real-time electrical brain activity noninvasively.
Data interpretation	Interprets emotional and cognitive states such as fatigue, distraction, or anxiety using EEG signal features.
Stimulation modalities	Uses sound waves, flashing lights, or low-level electrical currents to modulate brainwave patterns.
Adaptive mechanism	Machine learning algorithms adapt stimulation parameters in real time based on brain activity.
Closed-loop feedback	Continuously monitors brain response and adjusts stimulation accordingly, creating a responsive feedback loop.
Applications	Enhances learning, memory, focus, stress management, emotional regulation, VR immersion, and mental health treatment.

relevant brain activity. Continuous feedback allows for a closed loop of data to continuously adapt and tune the stimulation pattern and form a more dynamic responsive system to personal user needs in real time. The previously mentioned neurotechnology has many applications in improved learning, memory, focus, stress control, emotional regulation, VR immersion, and mental health, demonstrating the extensive nature of the technology for mental enhancement and therapeutic applications. This adaptive biofeedback system is versatile enough to approach a multitude of applications from personalized cognitive training and educational enhancement to VR immersion, mental health treatment, and stress management.

Also Figure 8.5 shows the operational system behind an Adaptive Closed-Loop EEG-Based Brain Enhancement System, which senses and reacts to brain activity in real time. The system starts with sensors that externally collect electrical signal information from the brain as the EEG signals. After recording the EEG signals, they are then subject to a real-time signal processing sequence to clean and prepare the channel signals. Specific signal channels are sent to a mental state analysis level for identifying the client's mental state/fatigue level. The ML module or decision engine based on the mental condition interpretation shapes the appropriate action signal. Next, the system provides stimulus output signals with adaptive stimulation based on the user's brain activity to modulate mental states and improve mental performance. Finally, the system also receives feedback from the user's brain in the form of brain response, completing the function of the closed-loop brain-enhancing system. The feedback mechanism allows for ongoing systematic adaptive stimulation of the brain that leads to personalized, optimized brain enhancement regardless of the user's initial state or background.

The invention in [37] is a clever, secure, and simple-to-use approach to recognize humans via their unique brain waves and the tendency of their brains to identify their passwords. The working principle behind it is a display, whether that be a screen, monitor, or projection that presents a set of different images, where each image is "flickering" at different frequencies. When the identifier looks at their "password" image, their brain gets a tendency to create one-of-a-kind brain waves at that instance, which are further detected via a cognitive interface, such as an EEG headset. The system checks the brain waves generated by the brain against the brain wave patterns stored in the database that were created when the right password image was identified previously. If the brain wave patterns match together, the system knows it is the "USER" and it lets you in; in case the brain wave patterns do not match, it does not permit the intruder in. The images supplied could be different each time the user logs in to make this advanced system even more secure. With the mix of a distinct image and the brain signals, it is absolutely difficult for any intruder to hijack your identity, and therefore this reliable system offers you excellent security for your accounts and devices.

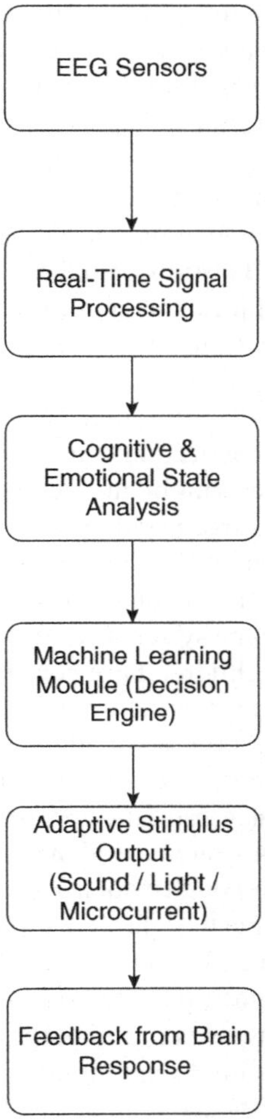

FIGURE 8.5 Adaptive closed-loop EEG-based brain enhancement system.

8.4 FINGERPRINT-BASED AUTHENTICATION MODEL

The research in [38] has demonstrated a new authentication approach that provides improvement over conventional biometric methods because it employs a sequence of multi-fingerprints when a user authenticates, rather than a traditional single fingerprint or password. The new proposal adds a biometric form of identification and a knowledge-based component by incorporating how the fingerprints are presented in sequence. Therefore, the novel method provided improved security and usability by minimizing the shortcomings of current approaches such as password reuse, memory intrusion, substantial storage of passwords, brute force, and phishing attacks, among many others. The study assessed the new system in comparison with a legacy model called BioPass, referring to biometrics with textual password. The new system prompts a user to register three fingerprints to use when registering and authenticating in a predetermined sequence. To authenticate as a user, a user has to replicate the authenticated fingerprints in the same sequence they were registered. The result is a system that is more resilient to unwanted intrusion even where unauthorized fingerprints are introduced. Using the widely used model created by Bonneau et al., the system was evaluated based on the usability and security criteria. The new system scored better than BioPass in measures where it matters: memory effort, scalability, physical ease, and error management. The new system was more secure than BioPass and provided greater security against unthrottled guessing and phishing. The comparative analysis of the study also revealed some of the same limitations as BioPass, with targeted impersonation and internal observation as relevant weaknesses, among others. Compared to BioPass at 143.8 and 124.3 ms, experimental confirmation with MATLAB also showed that our proposed system is more efficient with a mean registration time of 102.8 ms and mean authentication time of 84.9 ms (authentication). Since efficiency can lead to better user experience and even more security; our computer programming demonstrates a directional improvement in biometric authentication as well as a reduced memory load on users, a superior response time, and a difficult-to-duplicate sequence. These features provide a legitimate alternative, tightening the user's authentication to a non-conventional biometric-password model.

The study [39] investigates how fingerprint biometrics can identify people from very early childhood to adulthood. Researchers utilized a large national identity database from Uruguay that contained well over 200,000 fingerprint images of about 134,000 people aged from newborns up to 18 years. The primary aim was to determine if fingerprints obtained from children, and more critically, very young children, can still be recognized by matching systems generally developed for the adult population. An obstacle mentioned in the study was that children have considerably smaller fingers and that their fingerprint images don't have much development or definitions of the ridge patterns found in adults. Thus, the research was able to utilize a scaling process to adjust children's fingerprint images to a larger, adult-like representation. This scaling process was based on how fingerprint ridges develop in size over an age duration. After scaling, the fingerprint images were processed for pattern deposit detection using standard processing and recognition systems used in most fingerprint matching applications. The results were very encouraging. There was enough evidence to suggest that even one-year-old fingerprint images could be used for positive identification purposes almost as accurately as adult fingerprints, especially in the case of the scaling process. In addition, as fingerprint images were preprocessed, the quality and size of minutiae increased significantly. In conclusion, the research presented in this paper establishes that with a few modifications, existing fingerprint systems can successfully be used to identify children and even infants. This is important for national identity programs, vaccination monitoring, and child protection systems as early and reliable identification is essential. The support for the research suggests that biometric systems can be deployed at an early age, whether through admission to formalized systems or through systems utilized in informal environments. The research will inform future systems and policy to ensure thoughtful consideration of the biometric identification of everyone from babies to children to adults and the elderly.

Yu et al. [40] give an in-depth review of the literature on fingerprint sensor technologies, including their working principles, materials, performance characteristics, and area of use. It distinguishes between fingerprint sensors based on three standard types: optical, capacitive, and ultrasonic. Optical sensors are distinguished for their simple structure and low material cost, but are limited in environments that exhibit surface contamination. Capacitive sensors can be miniaturized and consume very little power, but do not have the same resolution as optical sensors. Ultrasonic sensors are superior for 3D imaging and are able to operate in the presence of moisture or dirt, making them the best option for smartphones and insecure access. The literature review also discusses fabrication methods, including Complementary Metal-Oxide-Semiconductor (CMOS) integration and MEMS-based (micro-electro-mechanical systems) processes, and their potential to enhance sensitivity, scalability, and potential for production. The review also compares the metrics that are commonly used to assess fingerprint sensor performance characteristics such as image resolution, response time, false acceptance/error proportion, power usage, and robustness. The application of all these commercial sensors ranges from mobile platforms, medical wearables, smart locks, and applications for border control. Additionally, the literature review discussed technology development such as combined technologies for hybrid sensors, transparent and flexible sensors for next-generation wearables, and the possible integration of AI to assist with recognition of unique patterns. Lastly, the paper does a great job summarizing the challenges faced with fingerprint sensors with respect to standardization, data privacy, and uncertainties about fabrication costs as well as real-world performance. To summarize, the review concludes all ultrasonic sensors are the most advanced, all capacitive the most scalable and old technology, and all optical the most established.

The paper [41] has provided a comprehensive privacy-preserving fingerprint authentication system that leverages fully homomorphic encryption (FHE) to keep fingerprint data encrypted throughout storage, processing, and matching. Unlike conventional methods that keep plaintext fingerprint templates or insecure hashes, the system uses encryption for fingerprints, encrypts FingerCode vectors, and performs operations in the encrypted domain—floating concentric circles, all while encrypting the computation and measuring the distances, without downgrading the accuracy. The system extended the Fast Fully Homomorphic Encryption over the Torus (TFHE) library to do arithmetic on encrypted double-precision floating-point numbers. A major technical contribution of the paper is the system's ability to compute the distance between encrypted vectors, without needing to quantize the floating-point representation of its sensors, which was unlike the previously offered solutions with partial homomorphic encryption on integer representations that ultimately suffered loss in accuracy. The system has the capability to tie-in (and connect) with the MySQL RDBMS (relational database management system) through a proxy server to enable SQL queries over encrypted data. In this case, encryption was managed with a key on the client side and the server was providing the evaluation key allowing evaluation, but as an "honest-but-curious" server, the outputs could not be decrypted. The system was evaluated using NIST Special Database 9, which contains 4,000 fingerprint samples to verify performance (or speed). The system was able to complete the matching on average in 166 seconds, while victim tests indicated EER (a validity threshold for matching biometrics) calculated with a low range and low stability across configurations that were deployed throughout the evaluations. The performance of TFHE was dramatically improved over Fully Homomorphic Encryption (FHE), particularly as either vector sizes or bit precision increased. The authors describe the storage and calculation overheads imposed by encrypting and inserting large vectors into the database as significant; still, considering the level of privacy provided, they believe this was a welcome trade-off. In summary, this work describes a method for accurate, privacy-preserving biometric authentication under FHE. It is particularly well aligned for sensitive secure web or cloud services. The authors do remark that future implementations could be even better and more practical with parallel computation and the use of adaptive thresholds instead of static, hard provisions to reduce computation time.

The systematic review in [42] assesses new developments in the field of fingerprint liveness detection (FLD) techniques that are developed to combat biometric systems against presentation attacks including those conducted via artificial fingers. This study reviews 27 high-quality studies that were completed during the years 2022 and 2023. Through this systematic review, the review identifies some notable advancements, including ML-based approaches, particularly using adversarial CNNs and lightweight transformer-based models like LFLDNet. Advances in IoT-enabled authentication systems that use multimodal approaches—liveness detection models were specifically looked at. The paper also highlights advancements in material-based detection techniques to identify and differentiate fake or downright spoof attempts by means of different materials such as latex, silicone, and wood glue.

Figure 8.6 shows the operational system behind an Adaptive Closed-Loop EEG-Based Brain Enhancement System, which senses and reacts to brain activity in real time. The system starts with sensors that externally collect electrical signal information from the brain as the EEG signals. After

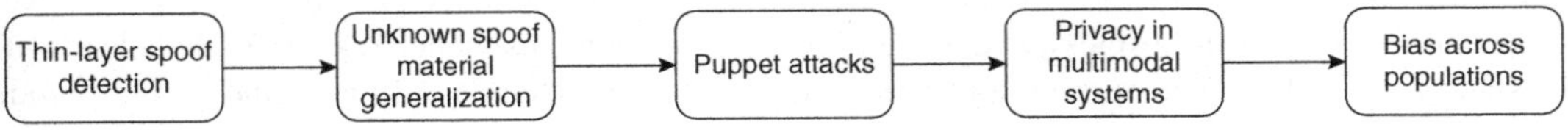

FIGURE 8.6 Key challenges in FLD.

recording the EEG signals, they are then subject to a real-time signal processing sequence to clean and prepare the channel signals. Specific signal channels are sent to a mental state analysis level for identifying the client's mental state/fatigue level. The ML module or decision engine based on the mental condition interpretation shapes the appropriate action signal. Next the system provides stimulus output signals with adaptive stimulation based on the user's brain activity, for modulating brain activity for better mental performance; finally, the system also receives feedback from the user's brain in the form of brain response, completing the function of the closed-loop brain-enhancing system. The feedback mechanism allows for ongoing systematic adaptive stimulation of the brain that leads to personalized, optimized brain enhancement regardless of the user's initial state or background.

Despite various advancements in the FLD community, the paper highlights various remaining challenges, e.g., detecting fake or thin-layer/subtle spoofs, the lack of research on generalizability across unseen spoof material, the threats posed by puppet-attacks, and potential privacy concerns with multimodal systems. In addition, the paper discusses some of the ethical concerns associated with algorithmic bias and fairness with regard to FLD models that were tested/deployed on different populations. The authors emphasized the need for improving accuracy with regard to fake detection while being considerate of user privacy and system transparency. Finally, the review stresses the need for future work to focus on cross-material generalization and privacy-preserving architectures, as well as making decisions that are ethically sound and adaptive in nature to support high-security areas such as banking, border control, and law enforcement.

Dwivedi et al. [43] discuss a crypto-biometric system that fuses fingerprint biometrics and cryptography to achieve secure two-party communication. Traditional biometric security systems are difficult to spoof and easy to steal—if they are obtained. On the other hand, independent cryptographic systems are also stolen events—independent cryptographic managers can be stolen as well—as is any encrypted data. This new paper overcomes some of these problems by proposing a hybrid model that uses fingerprint data to generate cryptographic keys. This ultimately ensures that only the "right" user could recreate the encryption key. The cryptographic keys are created by generating minutiae from the fingerprint image; then they are morphed into a binary string easy to work as a private cryptographic key. They use many techniques to achieve privacy (helper data generation), and more specifically unclear commitment for fingerprints. These allow slight variations during alternative runs of the same fingerprint scan, but with the same underlying minutiae in each scan—and thus the system should produce the same key. It is important for not only the privacy of user but also in these cases, the fingerprint itself cannot be restored with respect to generated feature minutiae and helper data—meaning that no fingerprint image can ever be reconstructed, since you alter the fingerprint minutiae and convert it into a binary string. They now can evaluate how well it behaves across multiple environmental variables to ensure that as a biometric security system it is accurate, robust, and easily impersonated.

Figure 8.7 represents the architecture of the Crypto-Biometric Model that utilizes fingerprint biometrics with cryptographic security for secure data communication. The operation starts from the fingerprint sensor, which accepts the biometric input. It then goes into minutiae extraction, where the minutiae of the fingerprint are identified. Once the unique features are identified, they are used to creating a feature vector, which is to extract a binary key for secure processing. Through the unclear commitment scheme, we are able to produce helper data that accounts for variability in biometric inputs while maintaining overall security. In the end, the binary key is produced, which is converted into a cryptographic key for secure communication. This key will encrypt and decrypt information, ensuring that information remains secure and confidential to authenticated users. This architecture combines the uniqueness of biometric characteristics with cryptographic methods to enhance the ability to protect data. Ultimately, we find that the system was successful across a number of tests and results showed high levels of accuracy, robustness, and preventing spoofing incidents, which resoundingly helped with key stability as well as measurement of security strength. The system is evaluated under many conditions and shows low accuracy and high robustness and spoofing attack resistance. System performance

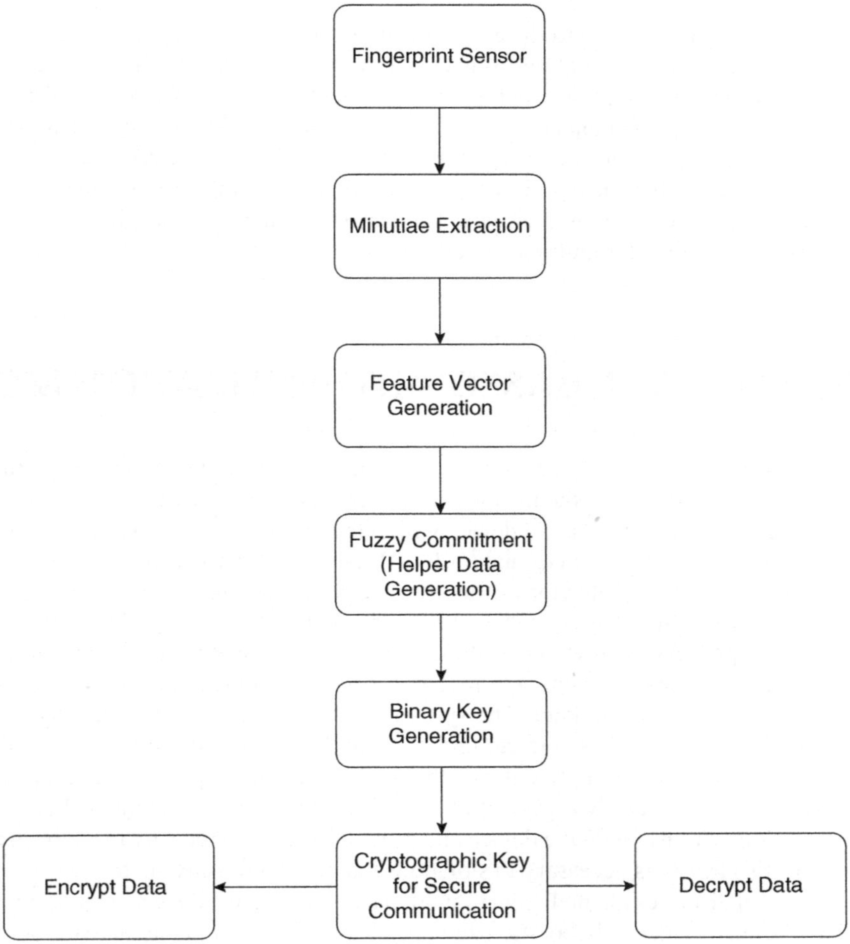

FIGURE 8.7 System architecture of the crypto-biometric model.

is also acceptable on key stability and security strength. Ultimately, results show that biometrics and cryptography, when combined, appear as a viable option to increase security in personal authentication and secure data-sharing applications, including personal banking, mobile device access, and confidential communications.

The paper [44] has considered and reviewed the problems of traditional identity authentication systems. They have proposed a hybrid model of an MD5-based cryptographic hash function and fingerprint recognition, which has been referred to as an Authentication Layer Cake. This dual-layer approach of biometric iris uniqueness and data encrypted with a hashing method improves on single-factor authentication methods. There are a number of notable limitations within the study. The limitations stem from the reliance on MD5, which is a cryptographic hash function that is vulnerable to collision attacks. If devices that require security are to be paired with the authentication, the strength of the cryptographic layer is questionable when you consider not only the capabilities of the MD5 itself, but also the other factors associated with limitations of hashing. Most literature has stopped using MD5; instead, more secure hashing algorithms such as SHA-256 or bcrypt are now being widely accepted. The vulnerability of MD5 does not appear to provide additional benefits for even one factor, let alone two-factor identification methods. While they introduced a second factor of fingerprint recognition, there was no mention of liveness detection or spoofing mitigation like prevention of synthetic fingerprint forgery. Authors find supportive false acceptance and rejection incident rates, as they have tested using a limited number of users with fingerprint templates, which lowers generalizability and statistical importance of a few tests of a small group of subjects. Elements related to usability were recorded for time for registration and time for login, but these were not compared with existing systems; therefore, it detracts from the overall comparative nature of the study. The paper introduces a basic prototype with practical utility and recognizes the need for multiple factors of authentication.

8.5 BEHAVIORAL-BASED AUTHENTICATION MODEL

The paper [45] discusses how the invention relates to an intelligent electronic device that supports secure, seamless, and continuous authentication of a user. Rather than requiring only the traditional login and authentication mechanisms, the device will authenticate using real-time aspects of the user and how the user interacts with the device naturally—through their own typing. As the user types, the device will monitor the distinct behavior of the user behavioral biometrics through the user's typing ability, which varies from user to user, such as the user's typing patterns, typing rhythm, and timing between characters typed. As the user types, the device will capture the user's behavioral biometrics, which will enhance the device's resolution to recognize the environment, and will extract the defined attributes as well as raw timestamp data. The system will use a trained NN to learn the user typing behavior, and on a case-by-case basis, if the user's typing behavior was similar to the user's typing behavior, the device will note that it recognizes the user and continue performing recognition in the background while the user naturally types with the device. This provides heightened security without disrupting the user's experience with the device and provides seamless security in environments where monitoring the user's identity is necessary in situations that handle sensitive data.

The patent in [46] discloses an intelligent system that acts to change the way people collaboratively work "whether it be chat", video collaboration, or collaborating in a shared interactive workspace based on their cognitive readiness to collaborate. The system contains an intelligent engine, which has been labeled the "collaboration mode transition engine", to evaluate the person's biometrics such as heart rate, brain waves, and behavioral, which include typing speed and attention span, so that the system can evaluate the person's cognitive state. The collaboration mode transition engine will evaluate this cognitive state and compare it with the cognitive threshold to the person's current collaboration mode. The collaboration mode transition engine will determine the person's mental operating level is greater than the cognitive threshold that is required by their collaboration mode, and automatically move this user to an equally more challenging/immersive collaboration mode. This creates a much more personalized, productive, and reactive digital collaboration experience based upon the real-time condition of how the user is feeling.

This invention mentioned in [47] relates to an electronic device that authenticates a user and continually reconfirms their identity from their human behavioral biometrics, their unique typing behavior. The device contains a memory device storing program instructions to be executed by a processor to avoid confusion about the order of program instructions such as running several programs simultaneously. When a user starts a session on the device, their typing behavior is constantly measured in real time while they type. This data is then processed to create measures such as speed, rhythm, and time between keys pressed, and it creates a timestamp with that behavior. These timestamps can then be analyzed with a trained NN model, to produce timestamps specific to user preference; if it fits, the user has been authenticated and is relevant to their profile. Unlike traditional one-off methods of authentication, this device will observe the user while they are using the device and authenticate continuously on a drive; this enhances security by ensuring that the person who started the session is still the same person using the device.

The paper [48] talks about how the process relies on how a person generally works on the device, such as how they type or how they move. The intent is to recognize a person using their natural user behavior to securely authenticate them. When the user first sets up their profile, the system records the user's normal behavior and breaks the secret access code down, linking it to similar behaviors and develops a type of user's instructions to guide future code recovery. When the user attempts to log back in, the device observes the user for a detailed understanding of their activities, applies the context to piece their activities back together, and compares the reconstructed code with the original code. This allows the device to determine that this is indeed the same user, even though there is no "password" entered by the user.

Figure 8.8 shows the components of a Behavioral Biometric Authentication System that authenticates user identity based on the unique way they behave, such as typing rhythm, or patterns of violation. The user setup is the initial phase. The second phase centers on collecting behavioral data, which describes the user's input, such as their keystrokes, mouse activity, gestures, etc. The behavioral input is then subject to feature extraction to identify distinguishing features, or behavioral traits, which we can then use to create a behavioral profile. The behavioral profile is processed by completing the secret code breakdown and linking. This will be cached as the reference signature by which we authenticate the user. During user login, the system will undertake real-time observation or input of the user's behavioral data and match it against the previously stored profile in a secret coded manner. The last step involves the reconstruction of the original code, determining similarity against the previous behavioral profile, and deciding whether to accept or reject the login attempt. This method of authentication is continuous, does not require a model of authentication which the users must manage past the initial setup, and is unique to each user's behavior.

Araújo et al. [49] provide a static keystroke dynamics-based user authentication system that builds on traditional password security by investigating how the user types utilizing keystroke captures. The system identifies four key biometric facts from keystroke data: ASCII key code, Down-Down (DD) latencies, Up-Down (UD) latencies, and Down-Up (DU) latencies. These values are taken as the user types a predetermined passphrase string and provides the reference base to develop the user's template. With the

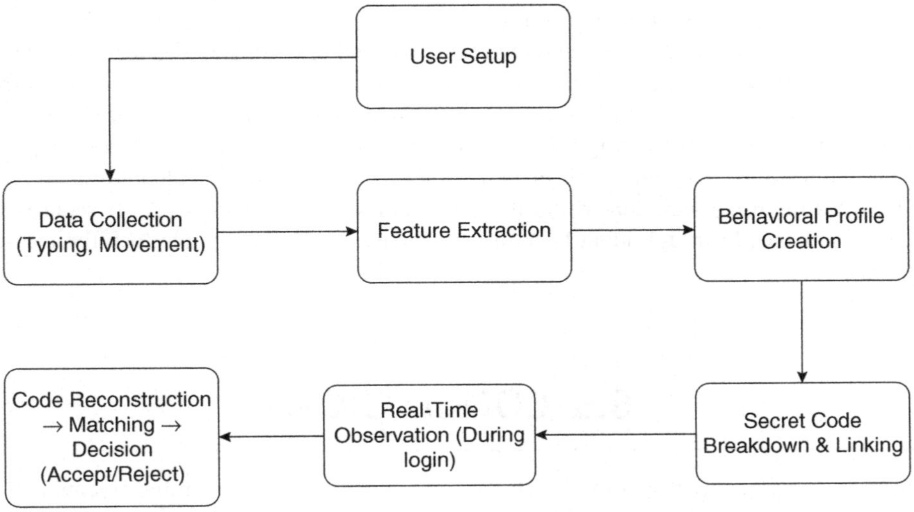

FIGURE 8.8 Workflow of behavioral biometric authentication.

TABLE 8.11 Trust score

TRUST SCORE RANGE	*SYSTEM ACTION*	*EXAMPLE BEHAVIOR*
80–100 (high)	Grant access	Normal typing, known context
50–79 (medium)	Break down behavior → re-evaluate	Fast typing in new location
0–49 (low)	Escalate to assistive processing	Unknown app usage, erratic input pattern

user's template, an analogous input to authenticate is then passed through statistical distance thresholds to compare its likeness using the same biometric features. The study involved three types of users: a genuine user, a direct impostor, and an observer impostor. Each participant provided a set of standard samples with a data set of 30 users taking 10 attempts each. The study results showed that all four features together provided the best performance with a 1.45% FRR and a 1.89% False Acceptance Rate (FAR). The system also demonstrated good resistance against observer impostors who had only observed the authentic users typing the password. Other contributions of the work included examining high-resolution timing, allowing for controlled types of enrollment samples, and the optional template adaptation to counter long-term deterioration in performance. Compared to the earlier studies, the final system design provided good comparative performance with no extra requirements for special hardware.

Burström et al. [50] mention how behavioral security becomes enhanced by the device not only understanding how a user interacts with the device but also in the situation or context in which the user interacts with the device and what they are doing when they interact. It will give an explanation if we stop at what a user was doing when they interacted and viewed their captured "behavior" in real time and gave it a "trust score" based on that interaction. If the interaction scored low on trust, rather than make a value judgment like "the user is acting strange", it breaks the behavior down into smaller parts based on what the user was doing when they interacted. After enough of the context-specific behavior data is used to develop a second, more accurate "trust score" for the behavior, if the second score is still somewhat low, then the device communicates to the rest of the system to assist in processing in order to help catch irregular or risky behavior as precisely and fairly as it can. It is indicated in Table 8.11.

Table 9.5 depicts system responses according to trust scores which likely indicate a confidence level in the user's legitimacy or behaviors. Trust scores are organized into three ranges: High (80–100), Medium (50–79), and Low (0–49). The High trust score assumes the user is a legitimate high use, the system gives access with no interruptions based on typical typing patterns in a familiar context. Medium is where the system identifies the behavior and re-evaluates the scenario based on the user's potential abnormality. A user typing abnormally fast or accessing content from an unfamiliar geographic location could all be indications of a potential abnormality. Low trust score responses lead to upgrade processing, indicating possible suspicious behavior such as abnormal input and use of unknown applications. This approach of multi-level trust responses is valuable for the system as it allows for dynamic use simulation within the context of security principles. User security and user experience are enhanced through identifying user behaviors while being able to fully accommodate the appropriate use.

8.6 CONCLUSION

Smart buildings are gradually becoming interlinked and intelligent; it is becoming essential as an adaptive and resilient security infrastructure. This chapter extensively explored the development of biometric authentication technology with an emphasis on investigating EEG-based brainwave signals, behavioral biometrics, and fingerprint. It was prominent in the chapter that biometric authentication systems are evolving from single-factor authentication to multi-factor systems with adaptive identity verification capabilities. Synthesis of published literature with recent patents also suggests that there is a multitude of potential multi-layered biometric systems for meeting the security with the smart grids' increasingly complex security requirements. The improvements being developed in the space of biometrics do not only provide more reliable authentication but also open the door to user-based intelligent security frameworks.

REFERENCES

1. Chaudhary, P. and Agrawal, R. 2018. Emerging threats to security and privacy in brain computer interface. *International Journal of Advanced Studies of Scientific Research*, 3(12). https://papers.ssrn.com/sol3/papers.cfm?abstract_id=3326692.
2. Zhang, S., Sun, L., Mao, X., Hu, C. and Liu, P. 2021. Review on EEG-based authentication technology. *Computational Intelligence and Neuroscience*, 2021(1), p. 5229576. doi:10.1155/2021/5229576.
3. Chaurasia, A.K., Fallahi, M., Strufe, T., Terhörst, P. and Cabarcos, P.A. 2024. NeuroIDBench: An open-source benchmark framework for the standardization of methodology in brainwave-based authentication research. *Journal of Information Security and Applications*, 85, p.103832. https://www.sciencedirect.com/science/article/pii/S2214212624001340.
4. Isa, I.S., Zainuddin, B.S., Hussain, Z. and Sulaiman, S.N. 2014. Preliminary study on analyzing EEG alpha brainwave signal activities based on visual stimulation. *Procedia Computer Science*, 42, pp. 85–92. https://pdf.sciencedirectassets.com/280203.
5. Priyadarshani, M., Kumar, P., Babulal, K.S., Rajput, D.S. and Patel, H. 2024. Human brain waves study using EEG and deep learning for emotion recognition. *IEEE Access*. https://ieeexplore.ieee.org/stamp/stamp.jsp?arnumber=10597545.
6. Dutta, P., Paul, S., Cengiz, K., Anand, R. and Kumar, A. 2023. A predictive method for emotional sentiment analysis by deep learning from EEG of brainwave dataset. In: R. K. Kalita and S. Sharma (Eds.), Artificial Intelligence for Neurological Disorders, pp. 25–48. Academic Press. doi:10.1016/B978-0-323-90277-9.00002-X.
7. Nordin, N.F. and Alias, N. 2022. The classification of human emotions based on the electroencephalogram (EEG) of brain waves. *Proceedings in Science and Mathematics*, 9, pp. 286–297. https://science.utm.my/procscimath/wp-content/uploads/sites/605/2022/10/29.-NABILAH-FARHANA-BINTI-NORDIN-A18SC0159-1.pdf.
8. Li, X., Zhang, Y., Tiwari, P., Song, D., Hu, B., Yang, M., Zhao, Z., Kumar, N. and Marttinen, P. 2022. EEG based emotion recognition: A tutorial and review. *ACM Computing Surveys*, 55(4), pp. 1–57. https://doi.org/10.1145/3524499.
9. May, T.J., Joyce, A.C. and Briana, B. 2022. Facilitating payments in an extended reality system based on behavioral biometrics. U.S. Patent 17/548350. https://patents.google.com/patent/US20220188833A1/en.
10. Revanth, D.S., Subhash, M.S., Kesav, M. and Reddy, B.H. 2024. Secure touch: IoT-enabled finger-touch authentication for seamless mobile security. *Journal of Microsystems and IoT*, 2(6), pp. 924–929. https://doi.org/10.5281/zenodo.13221623.
11. Palaniappan, R. 2008. Two-stage biometric authentication method using thought activity brain waves. *International Journal of Neural Systems*, 18(01), pp. 59–66. https://d1wqtxts1xzle7.cloudfront.net/31337569.
12. Soni, Y.S., Somani, S.B. and Shete, V.V. 2016. Biometric user authentication using brain waves. In: *2016 International Conference on Inventive Computation Technologies (ICICT)*, Coimbatore, India (Vol. 2, pp. 1–6). IEEE. https://d1wqtxts1xzle7.cloudfront.net/57216011.
13. Rahul, M., Sakshi, P., Archana, M. 2023. Authentication using behavioral biometrics system. https://curin.chitkara.edu.in/wp-content/uploads/2023/11/202311061348.pdf.
14. Schwartz, A. L., and Petrovic, V. 2019. Systems and methods for biometric recognition. U.S. Patent US20190122024A1.
15. Alipio, M. I. 2023. Development, evaluation, and analysis of biometric-based bank vault user authentication systems through brainwaves. *Journal of Ambient Intelligence and Humanized Computing*, 14(8), pp. 10165–10179. doi:10.1007/s12652-021-03679-8.

16. Arias-Cabarcos, P., Habrich, T., Becker, K., Becker, C. and Strufe, T. 2021. Inexpensive brainwave authentication: new techniques and insights on user acceptance. In: *The 30th USENIX Security Symposium (USENIX Security 21)*, Boston, MA, pp. 55–72. https://www.usenix.org/system/files/sec21-arias-cabarcos.pdf.
17. Koudelková, Z., Strmiska, M. and Jašek, R. 2018. Analysis of brain waves according to their frequency. *International Journal of Biological and Biomedical Engineering*, 12, pp. 202–207. https://www.researchgate.net/profile/Zuzana-Koudelkova-4/publication/334805116_Analysis_of_brain_waves_according_to_their_frequency/links/5d4194e84585153e59312c60/Analysis-of-brain-waves-according-to-their-frequency.pdf.
18. Fladby, K. 2008. Brain wave based authentication. Master's thesis. https://ntnuopen.ntnu.no/ntnu-xmlui/bitstream/handle/11250/143821/Kennet_Fladby_Brain_Wave_Based_Authentication_Final.pdf?sequence=1.
19. Fidas, C.A. and Lyras, D. 2023. A review of EEG-based user authentication: trends and future research directions. *IEEE Access*, 11, pp. 22917–22934. https://ieeexplore.ieee.org/document/10058960?denied=.
20. Li, G., Lee, C.H., Jung, J.J., Youn, Y.C. and Camacho, D. 2020. Deep learning for EEG data analytics: A survey. *Concurrency and Computation: Practice and Experience*, 32(18), p. e5199. https://www.kcdccollege.ac.in/admin/uploads/IMG-6586da530317d1.28959177.pdf.
21. Gui, Q., Jin, Z. and Xu, W. 2014. Exploring EEG-based biometrics for user identification and authentication. In: *2014 IEEE Signal Processing in Medicine and Biology Symposium (SPMB),* Antalya, Turkey, pp. 1–6. IEEE.
22. Shams, T.B., Hossain, M.S., Mahmud, M.F., Tehjib, M.S., Hossain, Z. and Pramanik, M.I. 2022. EEG-based biometric authentication using machine learning: A comprehensive survey. *ECTI Transactions on Electrical Engineering, Electronics, and Communications*, 20(2), pp. 225–241. https://ph02.tci-thaijo.org/index.php/ECTI-EEC/article/view/246906.
23. Hernández-Álvarez, L., Barbierato, E., Caputo, S., Mucchi, L. and Hernández Encinas, L. 2022. EEG authentication system based on one-and multi-class machine learning classifiers. *Sensors*, 23(1), p. 186. https://www.mdpi.com/1424-8220/23/1/186.
24. Tot, I., Trikoš, M., Bajčetić, J., Lalović, K. and Bogićević, D. 2021. Software platform for learning about brain wave acquisition and analysis. *Acta Polytechnica Hungarica*, 18(3), pp. 147–162. https://www.researchgate.net/profile/Tot-Ivan/publication/349934364_Software_Platform_for_Learning_about_Brain_Wave_Acquisition_and_Analysis/links/604a1a02a6fdcc4d3e5622f6/Software-Platform-for-Learning-about-Brain-Wave-Acquisition-and-Analysis.pdf.
25. Kopito, R., Haruvi, A., Brande-Eilat, N., Kalev, S., Kay, E. and Furman, D. 2021. Brain-based authentication: towards a scalable, commercial grade solution using noninvasive brain signals. bioRxiv, Art. no. 2021.04.09.439244v2, pp. 2021–2004. doi:10.1101/2021.04.09.439244v2.full.pdf.
26. Kuremoto, T., Ishikawa, J., Mabu, S. and Mitsushima, D. 2023. Recognition of brain wave related to the episode memory by deep learning methods. In: A. El hajj and R. Boutaba, (Eds.), Research Advances in Data Mining Techniques and Applications. IntechOpen. https://www.intechopen.com/chapters/87827.
27. Alkhyeli, Z., Alshehhi, A., Alhemeiri, M., Aldhanhani, S., AlBalushi, K., AlNuaimi, F.A. and Belkacem, A.N. 2022. Secure password using EEG-based brainprint system: unlock smartphone password using brain-computer interface technology. In: *2022 IEEE International Conference on Bioinformatics and Biomedicine (BIBM)*, Las Vegas, NV, USA, pp. 1982–1987. IEEE. https://www.researchgate.net/profile/Fatima-Alnuaimi-6/publication/366833432.
28. Yildiz Yagiz, C., and Knepper Lawrence, E. 2019. Method and apparatus for maintaining a secure head-mounted display session. U.S. Patent 10345902B1. https://patents.google.com/patent/US10345902B1/en.
29. Kumar, R. and Sharma, A. 2025. Multimodal EEG-based authentication system. World Intellectual Property Organization, WO2025109249A1. https://patentimages.storage.googleapis.com/82/c9/49/8a3ff345abda50/WO2025109249A1.
30. Yousefi, F. and Kolivand, H. 2021. A new solution to the brain state permanency for brain-based authentication methods. In: *2021 1st International Conference on Artificial Intelligence and Data Analytics (CAIDA)*, Liverpool, UK, pp. 69–72. IEEE. https://researchonline.ljmu.ac.uk/id/eprint/25512.
31. Adebola, W. 2025. Intelligent emotion detection: the role of machine learning algorithms. https://www.researchgate.net/profile/Victor-Ogunrinde/publication/390977833.
32. Abo-Zahhad, M., Ahmed, S.M. and Abbas, S.N. 2014. A novel biometric approach for human identification and verification using eye blinking signal. *IEEE Signal Processing Letters*, 22(7), pp. 876–880. https://d1wqtxts1xzle7.cloudfront.net/106925219.
33. Chandanwala, A.A., Bhowmik, S., Chaudhury, P. and Pravin, S.C. 2024. Hybrid quantum deep learning model for emotion detection using raw EEG signal analysis. https://arxiv.org/pdf/2411.17715.

34. Omar, W.R.W., Taib, M.N., Jailani, R., Fuad, N., Isa, R.M., Jahidin, A.H. and Sharif, Z. 2013. Acute ischemic stroke brainwave classification using relative power ratio cluster analysis. *Procedia-Social and Behavioral Sciences*, 97, pp. 546–552. https://pdf.sciencedirectassets.com/277811.
35. Pereira, D.R., Pereira, H.R., Silva, M.L., Pereira, P. and Ferreira, H.A. 2025. Impact of five basic tastes perception on neurophysiological response: Results from brain activity. *Food Quality and Preference*, p. 105572. https://www.sciencedirect.com/science/article/pii/S0950329325001478.
36. Poltorak, A. 2023. Method and apparatus for neuroenhancement to facilitate learning and performance. U.S. Patent 20230398356A1. https://patents.google.com/patent/US20230398356A1/en.
37. Agrafioti, F., Bui, F.M. and Hatzinakos, D., Nymi Inc. 2017. Enabling continuous or instantaneous identity recognition of a large group of people based on physiological biometric signals obtained from members of a small group of people. U.S. Patent 9646261. https://patentimages.storage.googleapis.com.
38. Algarni, M.H. 2024. Fingerprint sequencing: An authentication mechanism that integrates fingerprints and a knowledge-based methodology to promote security and usability. *Engineering, Technology & Applied Science Research*, 14(3), pp. 14233–14239. https://mail.etasr.com/index.php/ETASR/article/view/7250.
39. Preciozzi, J., Garella, G., Camacho, V., Franzoni, F., Di Martino, L., Carbajal, G. and Fernandez, A. 2020. Fingerprint biometrics from newborn to adult: A study from a national identity database system. *IEEE Transactions on Biometrics, Behavior, and Identity Science*, 2(1), pp. 68–79. https://www.researchgate.net/profile/Javier-Preciozzi/publication/338470962.
40. Yu, Y., Niu, Q., Li, X., Xue, J., Liu, W. and Lin, D. 2023. A review of fingerprint sensors: Mechanism, characteristics, and applications. *Micromachines*, 14(6), p. 1253. https://www.mdpi.com/2072-666X/14/6/1253.
41. Kim, T., Oh, Y. and Kim, H. 2020. Efficient privacy-preserving fingerprint-based authentication system using fully homomorphic encryption. *Security and Communication Networks*, 2020(1), p. 4195852. doi:10.1155/2020/4195852.
42. Ametefe, D.S., Sarnin, S.S., Ali, D.M., Muhamad, W.N.W., Ametefe, G.D., John, D. and Aliu, A.A. 2024. Enhancing fingerprint authentication: a systematic review of liveness detection methods against presentation attacks. *Journal of The Institution of Engineers (India): Series B*, 105(5), pp. 1451–1467. https://www.researchgate.net/profile/Divine-Ametefe/publication/380537324.
43. Dwivedi, R., Dey, S., Sharma, M.A. and Goel, A. 2020. A fingerprint based crypto-biometric system for secure communication. *Journal of Ambient Intelligence and Humanized Computing*, 11(4), pp. 1495–1509. https://arxiv.org/pdf/1805.08399.
44. Fujun, F., Xinshe, L. and Litao, W. 2016. Design and implementation of identity authentication system based on fingerprint recognition and cryptography. In: *2016 2nd IEEE International Conference on Computer and Communications (ICCC)*, Chengdu, China (pp. 254–257). IEEE. https://ieeexplore.ieee.org/abstract/document/7924704.
45. Sambamurthy, N.S. and Krishnan, P. 2017. *Security Device, Methods, and Systems for Continuous Authentication*. NSS Lab Works LLC. U.S. Patent 9852275 B2. https://patentimages.storage.googleapis.com/51/66/1c/59aadb27c324b6/US9852275.
46. Delgado, C. G., Rakshit, S. K., Dhillon, J. S., Lu, F. 2023. Collaboration mode transition based on cognitive overload. U.S. Patent 20250044868A1. https://patents.google.com/patent/US20250044868A1/en?oq=US%E2%80%AF2025%2f0044868%E2%80%AFA1.
47. Mishra, J. P. and Shukla, A. 2024. System, method and device for continuous user authentication and verification. WO2025012744A1. https://patents.google.com/patent/WO2025012744A1/en?oq=WO%E2%80%AF2025%2f012744%E2%80%AFA1.
48. Gelardi, G., Sajjad, A., and Ducatel, G. 2020. User authentication based on behavioral biometrics. U.S. Patent 12314362B2. https://patents.google.com/patent/US12314362B2/en.
49. Araújo, L.C., Sucupira, L.H., Lizarraga, M.G., Ling, L.L. and Yabu-Uti, J.B.T. 2005. User authentication through typing biometrics features. *IEEE Transactions on Signal Processing*, 53(2), pp. 851–855. https://ieeexplore.ieee.org/abstract/document/1381785.
50. Burström, P., Lindblad, P., Libell, T., Breitling, J. and Åhlén, M., LexisNexis Risk Solutions FL Inc. 2025. Systems and methods for improved behavioral scoring accuracy. U.S. Patent 12299094. https://patentimages.storage.googleapis.com.

Design and Implementation of a Quad-Mode ESP32-Powered Smart Home Automation System Driven by IoT

9

P. Pavithran, N. T. Bragadeeshwaran,
O. V. Gnana Swathika, and
K. T. M. Udayanga Hemapala

9.1 INTRODUCTION

9.1.1 Internet of Things

A. **Introduction**: A network of physical objects with sensors, software, and other technologies integrated to connect and share data with other devices via the internet is known as the Internet of Things (IoT). These gadgets can be anything from industrial machineries to home appliances. The primary objective of IoT is to enable real-time data collection, monitoring, and control in order to make the world smarter and more efficient. IoT makes it possible to automate and remotely manage systems, which increases performance, safety, and

DOI: 10.1201/9781003649700-9

convenience. For instance, smartphone can be used to control appliances in smart homes. IoT helps industries in monitoring equipment and minimize downtime.

B. **History**: The IoT origins can be traced back to the early years of the internet. When researchers at Carnegie Mellon University linked a Coca-Cola vending machine to the internet in the 1980s, they were able to verify that the drinks were cold and available. However, British technologist Kevin Ashton, who was employed at Procter & Gamble, is credited with officially introducing the term "Internet of Things" in 1999. He had an idea for a system that would use Radiofrequency Identification (RFID) to allow physical objects to speak to computers. The 2000s saw tremendous advancements in sensor technology and internet connectivity. Wider adoption was made possible by devices becoming more affordable, smaller, and energy-efficient. Numerous devices were able to connect to the internet after IPv6 was introduced.

9.1.2 Home Automation

A. **Introduction**: The term "home automation" describes the use of technology to automatically control and manage systems and appliances in the home. It enables users to control appliances like lights, fans, air conditioners, security cameras, and entertainment systems via voice commands, tablets, and smartphone. Improving convenience, security, energy efficiency, and general quality of life are the main objectives of home automation. Technologies like WiFi, Bluetooth, Zigbee, Z-Wave, and the IoT are essential to modern home automation systems. By enabling communication between devices and users, these technologies create a smart ecosystem in which appliances react intelligently to environmental changes or particular inputs.

B. **History**: When basic electrical appliances like vacuum cleaners, washing machines, and refrigerators were introduced in the early 20th century, they made household chores easier. This is when the history of home automation began. The development of X10 technology in the 1970s marked the first real step toward automation by enabling devices to communicate over electrical wiring to control appliances and lights. Home automation systems advanced in the 1980s and 1990s, but they were costly and primarily used in upper-class residences. Although they required intricate installations, they provided centralized control over security, climate, and lighting systems. The emergence of wireless connectivity and the internet in the 2000s created a huge change.

9.2 METHODOLOGY

The methodology used in this Smart Home Automation System is integrating day-to-day home appliances with IoT-based control logic in a modern way. The system design mainly focuses on flexibility, low cost, and multi-mode control with the help of ESP32 as a central controller. This approach first involves setting up the circuit by connecting the ESP32 with relay, sensors, and switches. Then the logic code is written, debugged, and later uploaded using Arduino IDE. Finally, the system is integrated with Arduino IoT Cloud allowing digital switch control through Google Home application, and voice control is added by linking Google Assistant to the system through cloud services like IFTTT (If This Then That).

9.2.1 Architecture

The architecture of this Smart Home Automation System is to control devices using multiple options

i. Manual control using switches
ii. Digital switch using Google Home
iii. IR Remote control
iv. Voice control with the help of Google Assistant

while maintaining seamless coordination through ESP32 microcontroller. The ESP32 acts as the central core of the system, which helps in receiving input signals and controlling output actions.

Four General Purpose Input Output (GPIO) pins of ESP32 are connected to four-channel relay module. The relay module is used to switch high-powered electrical appliances on and off. The DHT11 sensor is used in this system to monitor temperature and humidity. This sensor can help in potential environmental-based automation in future system enhancement.

For voice control, cloud-based services like Google Assistant and IFTTT are integrated. Whenever a voice command is given to Google Assistant or the digital switch is turned on the Arduino IoT Cloud creates an HTTP request to ESP32 via WiFi and the relevant relay reacts according to the command given.

This defined architecture includes multi-mode inputs which ensures high flexibility. This system allows real-time operation and future enhancement like mobile app integration, environmental-based automation, and sensor-based automation.

9.2.2 Hardware Description

The hardware setup of this Smart Home Automation System includes carefully selected components which enables reliable control of high-powered electrical appliances. Mainly five major standard components have been used in this system and its brief descriptions are listed below.

A. **ESP32 Dev Kit V1**: The ESP32 is a dual-core microcontroller with built-in Bluetooth and WiFi capabilities. It can perform as a complete stand-alone system or as a slave device to Microcontroller Unit (MCU), reducing communication stack overhead on the main processor. It acts as the central controller of the system, controlling input signals from the switches or remote control, and activates the appropriate relay channel after processing the signals. Its GPIO pins help in allowing multiple input and output connections simultaneously.
B. **4-Channel Relay Module**: A relay module is an electrically operated switch that uses a low-powered signal to control a high-power circuit. It acts as an intermediary by allowing a microcontroller or other low-power device to control higher voltage loads. Four GPIO pins of ESP32 are connected to individual relay channels, which enable the control of four different AC appliances. The relay module is activated using 3.3 V logic from ESP32.
C. **HX1838 IR Receiver Module**: The IR receiver module is used to receive the infrared signals from a remote control. This enables the wireless control of the relay by activating it only when the IR receiver module reads the correct hexadecimal code from the remote.
D. **DHT11 Temperature and Humidity Sensor**: This sensor monitors the surrounding temperature and humidity levels. It connects to the ESP32 and can provide real-time environmental data, which may be used for automated appliance control in the future system enhancements.

E. **Switches**: Standard PVC-Electrical switches are used in this system. These switches are connected to the ESP32 GPIO pins and act as a manual control to relays. This provides realistic and practical home implementation feel.
F. **Power Source**: The ESP32 and other modules need to be powered with a regulated 5 V DC power supply. Also, a 230 V AC mains power supply is needed for the electrical appliances connected with the relay module. The relay module prevents damages to ESP32 and other low-voltage components by isolating them from high-voltage loads.

9.2.3 Software Application

The platform used for developing logic code for Smart Home Automation System is Arduino IDE, which is used to write, compile, debug, and upload the code to the ESP32 microcontroller. The code implements logic for controlling the relays based on the inputs received from both physical and digital switches, remote control, and voice commands.

9.2.3.1 Programming environment

Arduino IDE is used for writing and uploading code to the ESP32. The logic code is written with the help of C++ programming language. The libraries used in the code are:

i. **ArduinoIoTCloud.h**: To enable Arduino devices to connect and interact with the Arduino cloud platform.
ii. **Arduino_ConnectionHandler.h**: To manage network connection for Arduino devices, enabling communication with the Arduino Cloud platform.
iii. **DHT.h**: For reading temperature and humidity values from DHT11 sensor.
iv. **IRremote.h**: For decoding IR signals from the HX1838 module.

9.2.3.2 Control logic

The software logic listens to four types of inputs which are manual control, IR Remote control, Digital switch control, and finally voice control.

i. **Manual Control**: Reads the state of GPIO pins connected to the switches. A HIGH state turns On the relay and the LOW state turns Off the switch.
ii. **Digital Switch Control**: Each appliance can be switched On/Off by tapping the virtual switch within the Google Home app interface shown in Figure 9.1.
iii. **IR Remote Control**: Infrared signals from the HX1838 receiver are decoded and only when appropriately hexadecimal code is read by the module the associated relay turns On/Off.
iv. **Voice Control**: Whenever a command is given, the IFTTT sends a webhook which requests ESP32 to turn On/Off the corresponding relay.

9.2.3.3 WiFi and IoT integration

The ESP32 connects with the WiFi network we provide, it can be home WiFi or even Mobile Hotspot and remains synchronized with the Arduino IoT Cloud. The commands from digital switch in Google Home or the voice control from Google Assistant is communicated to the ESP32 through the cloud. When the ESP32 receives the command it toggles the respective relay. The system enables real-time switching with minimal delay and supports multi-device control.

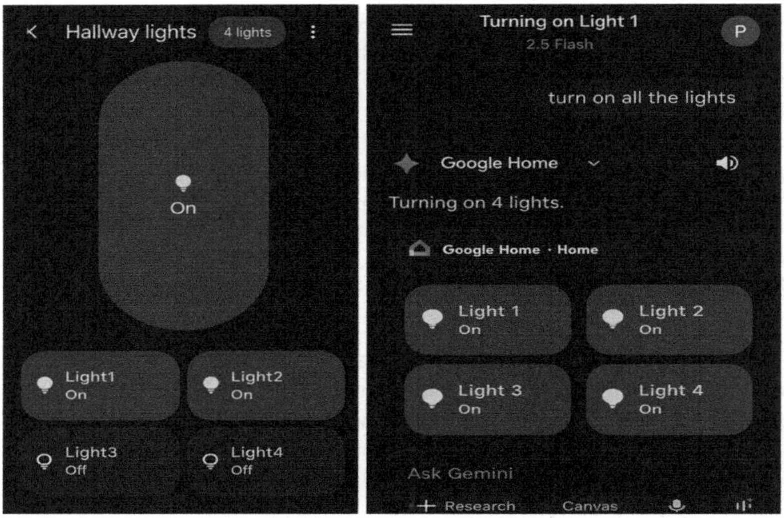

FIGURE 9.1 Google Home and Google Assistant interface.

9.2.3.4 Feedback and debugging

Serial communication is used for debugging purposes in Arduino IDE. The serial monitor option in Arduino IDE prints the relay status messages and sensor's reading. This helps in testing, finding the issues, debugging, and monitoring the system performances during the development stages.

9.2.4 Working Procedure

The working procedure of the Smart Home Automation System is built around the goal of providing users multiple flexible, synchronized ways to control electrical appliances within their home environment. Powered by the ESP32 microcontroller and integrated with Arduino IoT Cloud, the system combines four distinct control interfaces which are manual switches, IR remote, Google Home app, and Google Assistant voice commands, into one seamless and real-time automation framework. The overall workflow involves signal reception, processing, and execution via four-channel relays that switch the electrical appliances ON or OFF. The system also incorporates a DHT11 sensor for environmental temperature monitoring.

When the system is powered ON, the first step is the proper initialization of all components. A 5 V DC regulated power supply is provided to the ESP32 microcontroller, IR receiver, and DHT11 sensor, while the relay module and connected appliances are powered through 230 V AC supply. As soon as ESP32 boots, it initializes all its GPIO pins, configuring them as either inputs or outputs depending on their role. All four relays are initially set to a LOW state, ensuring that no appliances are unintentionally switched ON during the startup phase. Next, the ESP32 attempts to connect to the pre-configured WiFi network, and once successful, it establishes a secure link with the Arduino IoT Cloud. This ensures that remote and voice commands can be received in real time, and any changes to the system's state are synchronized across all control platforms.

After initialization, the system enters its main execution loop where it continuously listens for control signals from all four modes. The first and simplest control method is manual switch control. Standard PVC wall switches are connected directly to the ESP32's GPIO pins. When a switch is toggled it is detected by the microcontroller, which then processes the input and toggles the corresponding relay ON or OFF. This method ensures that the system remains fully operational even in the absence of an internet connection, making it practical and familiar for all types of users.

Simultaneously, the system is also listening for signals from the IR remote using the HX1838 IR receiver module. When a button on the remote is pressed, it sends an infrared signal containing a specific hexadecimal code. The IR receiver captures the signal, and the ESP32 decodes it using the IRremote.h library. If the received code matches one of the pre-assigned values in the program, the microcontroller activates or deactivates the corresponding relay channel. This enables offline wireless control and allows users to manage appliances from across the room without needing to physically interact with switches.

The third mode involves remote digital control through the Google Home app, made possible by connecting ESP32 with the Arduino IoT Cloud. When a user toggles the virtual switch for a specific appliance in the app, a cloud-based HTTP command is generated. This command travels through the internet and reaches the ESP32 in real time via WiFi. The microcontroller parses the command and triggers the correct GPIO output pin, causing the linked relay to switch ON or OFF. The updated relay status is then automatically pushed back to the cloud, ensuring the app reflects the true current state of the appliance, even if it was changed by a different control method.

The fourth and most user-friendly mode is voice control using Google Assistant. In this mode, when a user gives a voice command such as "Turn on fan" or "Switch off light 2," the command is first processed by Google Assistant and then routed through an IFTTT protocol. The IFTTT service is configured to send a webhook to the Arduino IoT Cloud, which in turn forwards the control signal to the ESP32 over WiFi. The ESP32 receives the command, determines the target relay, and toggles it accordingly. Voice control enhances accessibility for users who may be visually impaired or physically challenged, and also adds to the overall convenience and futuristic feel of the system.

Parallel to these control operations, the system also performs environmental monitoring using the DHT11 sensor, which is connected to one of the ESP32's GPIO pins. Every 10 seconds, the ESP32 reads the temperature and humidity values from the sensor and uploads them to the Arduino IoT Cloud as in Figure 9.2. These readings can be viewed in real time through the Google Home interface or can be known using voice commands like "What's the room temperature?" in Google Assistant. While the current implementation is focused on monitoring, the same data can be used for automated control in future upgrades. For example, we can turn on the fan if the temperature exceeds 30°C as in Figure 9.3.

A key aspect of the system's functionality is its real-time synchronization across all control modes. Whether a device is switched ON through a physical switch, IR remote, Google Home app, or voice

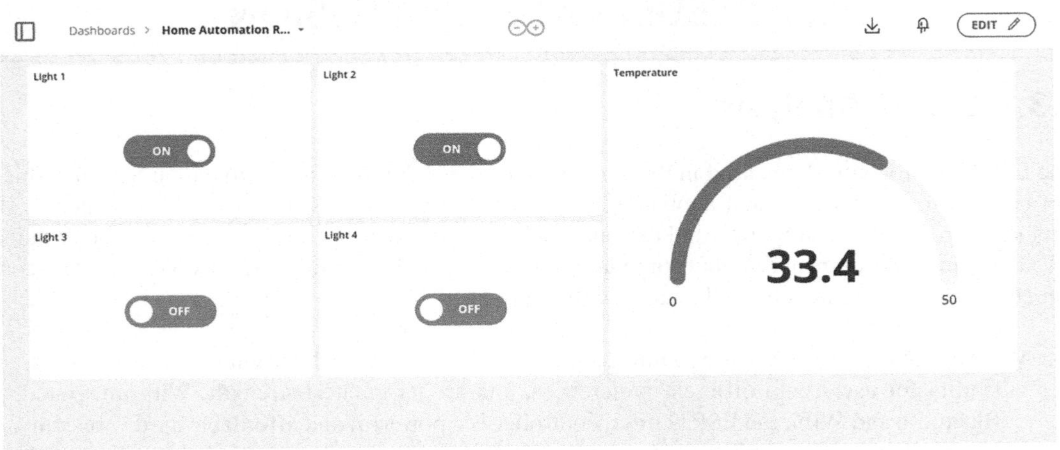

FIGURE 9.2 Cloud interface.

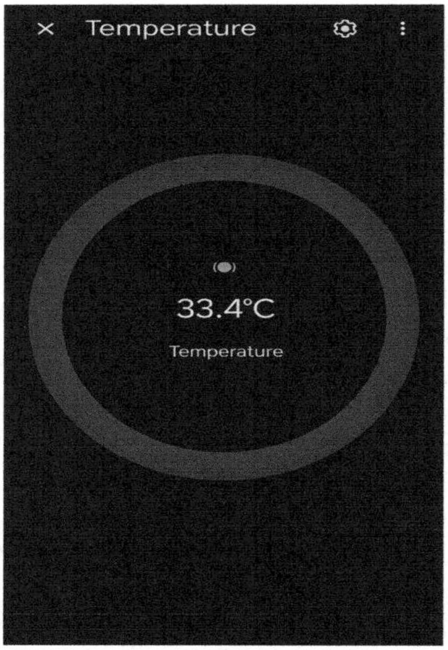

FIGURE 9.3 Temperature in Google Home application.

command, the ESP32 ensures that the updated status is reflected across all platforms. This prevents conflicts, confusion, or discrepancies about whether a device is ON or OFF.

Overall, our multi-mode system successfully integrates hardware and cloud services into a single responsive unit as in Figure 9.4. It supports instant reaction to user commands, maintains continuous connectivity with the cloud, and allows remote monitoring and control. The ESP32's built-in WiFi and processing power make it a perfect fit for this purpose, ensuring the smart home system remains reliable, efficient, and user-centric in daily operation.

9.3 RESULTS AND DISCUSSION

9.3.1 SWOT Analysis

The ESP32 microcontroller is used in the construction of this Smart Home Automation System, which can be used to control electrical appliances via a variety of interfaces, including PVC switches, IR remote controls, voice commands, and digital switches. It also includes remote access via the Arduino IoT Cloud and environmental monitoring via a DHT11 sensor. Flexibility, energy efficiency, and convenience in home automation are the goals of this system.

A. **Strengths**: The system uses multi-mode control interfaces, which guarantee easy accessibility for users with different preferences, and are its greatest strength. With integrated Bluetooth and WiFi, the ESP32 microcontroller is a powerful and affordable hardware component that lowers hardware complexity. User convenience is increased through smooth remote monitoring and control made possible by integration with the Arduino IoT Cloud.

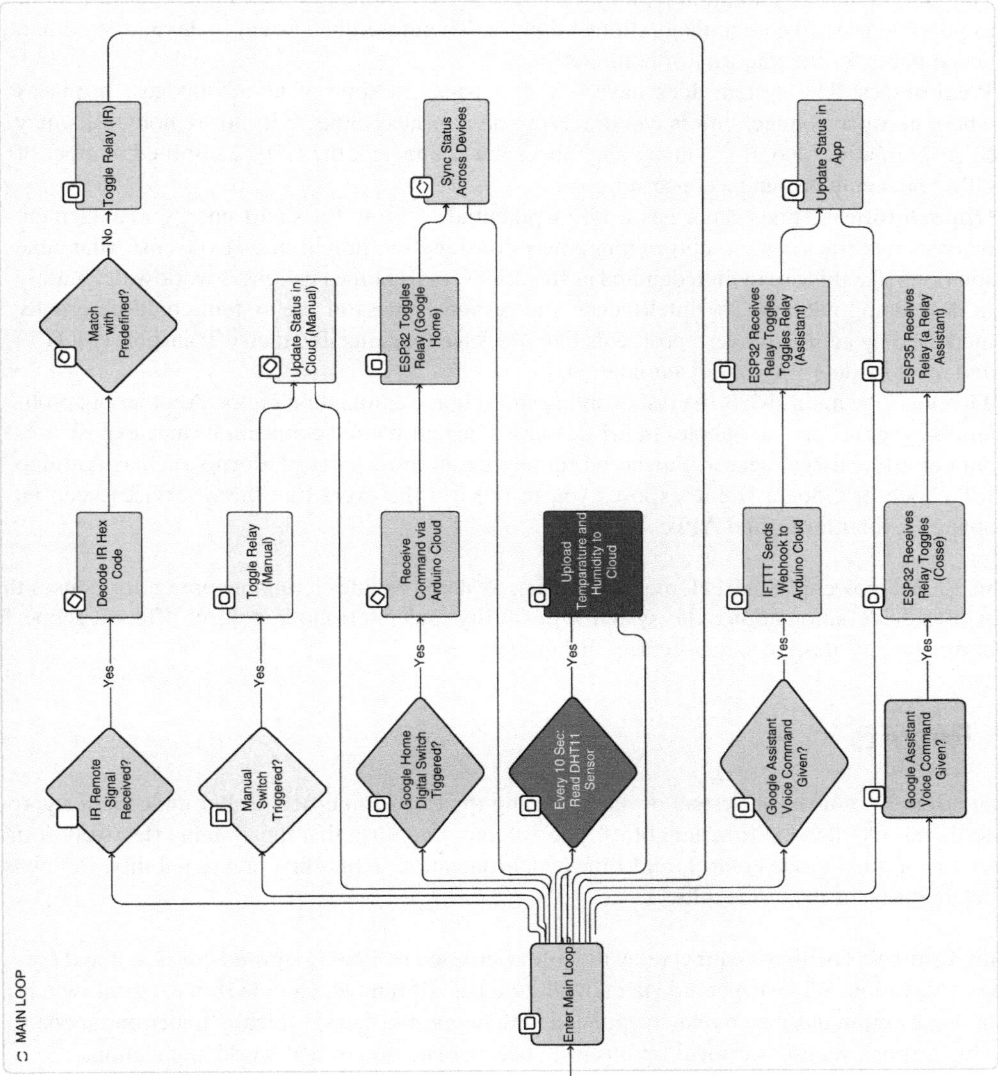

FIGURE 9.4 Working flow of a Smart Home Automation System.

The DHT11 sensor's addition raises awareness of environmental issues and opens the door to possible uses like automated climate control. Therefore, the system is flexible, modular, and suitable for the majority of home settings.

B. **Weaknesses**: The system does have few drawbacks in spite of its advantages. In places where network connectivity is worst, relying on internet connectivity for remote tasks may be problematic. Also, if too many appliances are connected, the ESP32's limited number of GPIO pins may become a concern.

C. **Opportunities**: This system has a lot of potential to grow for smart energy management features like tracking and optimizing power consumption in real time. Low-cost, adaptable solutions like this are in high demand as the use of smart homes increases worldwide, mainly in developing nations. The intelligence and responsiveness of the system could be greatly increased by adding modern protocols like Message Queuing Telemetry Transport (MQTT) and machine learning-based automation.

D. **Threats**: The main risk is the fast evolving smart home automation sector. Another big problem is cyber-security loopholes in IoT devices, since unsecured connections may expose user data or allow illegal access. Furthermore, relying on third-party platforms such as Arduino IoT Cloud or Google Home exposes you to risks in the event that these services keep on changing their terms and APIs.

Thus, this ESP32-powered Smart Home Automation System provides a reliable option for houses that looks for affordable automation. The system's flexibility and multi-mode control offers a scope for future improvements, despite some security difficulties.

9.3.2 Results

The Smart Home Automation System developed using the ESP32 microcontroller underwent rigorous testing across all its intended functionalities. The outcomes confirm that the system effectively fulfills the objectives of quad-mode control, real-time synchronization, reliability, and scalability. Below is a detailed elaboration of the test results:

Relay Control: The four-channel relay module controlled by ESP32 showed consistent and precise operation. When triggered via manual switches, IR remote, Google Home digital switch, or voice commands, each relay responded with negligible delay, typically under one second. This responsiveness is critical for ensuring user satisfaction in real-world applications.

IR Remote Functionality: The IR receiver module (HX1838) was successfully integrated and tested. It was able to detect and decode pre-programmed hexadecimal IR signals sent from the remote. Each recognized code reliably toggled the respective relay, demonstrating that users can conveniently operate appliances offline, without relying on Wi-Fi or internet connectivity.

Digital Control through Google Home: The ESP32 was seamlessly linked to the Arduino IoT Cloud and configured with Google Home for app-based control. Each virtual switch mapped to a physical relay could be toggled from the mobile application. State changes in appliances were reflected accurately and updated in real time, confirming stable cloud communication.

Voice Control through Google Assistant: The system's voice interface was configured using IFTTT webhooks and integrated with Google Assistant. Voice commands like "Turn on Light 1" or "Switch off Fan" were recognized quickly and converted into actionable HTTP requests. The ESP32 interpreted these requests and executed the associated relay operations efficiently. Voice control proved especially useful for accessibility and convenience.

Sensor Monitoring: The DHT11 sensor accurately captured environmental temperature and humidity readings and pushed this data to the Arduino IoT Cloud every 10 seconds. These readings were accessible via the cloud dashboard and potentially through smart assistant queries. The continuous flow of real-time sensor data opens avenues for future adaptive automation, such as switching on a fan if room temperature crosses a defined threshold.

Multi-Mode Synchronization: One of the system's standout features was its synchronization capability across all four control methods. Regardless of the control mode used to change a device's state, that change was reflected across the Google Home app, voice assistant, and physical interface. This eliminated state conflicts and ensured a coherent user experience.

System Stability: During prolonged testing, the system showed no signs of crashes or inconsistent behavior. The Arduino IDE's serial monitor validated that all components (relays, IR receiver, sensors, and cloud interface) operated as expected. This robustness underscores the platform's potential for daily real-world deployment.

User Experience: The multiple control options made the system accessible to a wide range of users from those who prefer traditional switches to those who rely on digital or voice-based interfaces. This inclusiveness enhances its usability for elderly users, individuals with disabilities, or tech-savvy users seeking automation.

In summary, the results demonstrate that the system is reliable, responsive, and easy to operate. It meets its core objectives while remaining scalable and adaptable for future upgrades, including energy monitoring and AI-driven automation.

9.4 CONCLUSION

This chapter presents a successful design and a working implementation of a low-cost, flexible, and user-friendly Smart Home Automation System using a powerful ESP32 microcontroller. This system perfectly enables control through multiple modes which includes manual switch control, digital switch control, IR remote control, and voice control. Integration of this system with Arduino IoT Cloud allows real-time access in monitoring the synchronization of all control modes.

The implementation of DHT11 sensor helps in knowing the surrounding environmental condition and also provides a room for future enhancements.

The main core of this system is ESP32 microcontroller which has a built-in WiFi and Bluetooth, making it a suitable platform source for this Smart Home Automation System.

To sum up, the project achieves all its goals and provides convenient, energy-efficient, and flexible system to the user. With few more minor advancements like MQTT protocol, environmental-based automation, energy monitoring, and machine learning-based automation, the system can be upgraded as a Fully Intelligent Smart Home Automation System.

BIBLIOGRAPHY

1. Sayeduzzaman, M., Hasan, T., Nasser, A.A. and Negi, A., 2024. An internet of things-integrated home automation with smart security system. In: *Automated Secure Computing for Next-Generation Systems*, A. K. Tyagi, Ed. Hoboken, NJ, USA: John Wiley & Sons, 2024, pp. 243–273.
2. Ajagbe, S.A., Adeaga, O.A., Alabi, O.O., Ikotun, A.B., Akintunde, M.A. and Adigun, M.O., 2024. Design and development of Arduino-based automation home system using the internet of things. *Indonesian Journal of Electrical Engineering and Computer Science*, 33(2), pp. 767–776.
3. Al Tareq, A., Mostofa, M.R., Rana, M.J. and Rahman, M.S., 2024. A comprehensive review of intelligent home automation systems using embedded devices and IoT. *Control Systems and Optimization Letters*, 2(2), pp. 198–203.
4. Adhikary, A., Halder, S., Bose, R., Panja, S., Halder, S., Pratihar, J. and Dey, A., 2024. Design and implementation of an IoT-based smart home automation system in real world scenario. *EAI Endorsed Transactions on Internet of Things*, 10, 6201. doi:10.4108/eetiot.6201.
5. Alturki, N., Alharthi, R., Umer, M., Saidani, O., Alshardan, A., Alhebshi, R.M., Alsubai, S. and Bashir, A.K., 2024. Efficient and secure IoT based smart home automation using multi-model learning and blockchain technology. *CMES-Computer Modeling in Engineering and Sciences*, 139(3), pp. 3387–3415.
6. Ahmed, I., Amjad, A. and Mehmood, M.A., 2024. Review paper on IoT based smart applications, home automation. *LC International Journal of Stem*, 5(1), pp. 45–58.
7. Thomas, L., MV, M.K., SL, S.D. and BS, P., 2024. Towards comprehensive home automation: leveraging the IoT, node-RED, and wireless sensor networks for enhanced control and connectivity. *Engineering Proceedings*, 59(1), p. 173.
8. Tay, S.W., Zhang, N. and AlJanah, S., 2024. A problem analysis of smart home automation: Toward secure and usable communication-based authorization. *IEEE Access*, 12, pp. 18103–18121.
9. de Boer, P.S., van Deursen, A.J. and van Rompay, T.J., 2024. The lights are on, but no one's home: A performance test to measure digital skills to use IoT home automation. *New Media & Society*, 26(9), pp.5259–5290.
10. Habib, M.R., Yusuf, M.A., Warnasuriya, W.N., Sunny, K., Rahaman, M.M., Khan, M.R.K., Saha, P.P. and Alam, M.T., 2024. A comprehensive review on the advancement of home automation system. In: *2024 Second International Conference on Intelligent Cyber Physical Systems and Internet of Things (ICoICI)*, August, Coimbatore, Tamil Nadu, pp. 638–642. IEEE.
11. Reddy, V.B., Dinesh, B., Manikyam, B., Gayatri, G., Kumar, S., Surekha, P. and Anand, S., 2024. Home automation using artificial intelligent & internet of things. *MATEC Web of Conferences*, 392, p. 01058.
12. Sanjay, C., Jahnavi, K. and Karanth, S., 2024. A secured deep learning based smart home automation system. *International Journal of Information Technology*, 16(8), pp.5239–5245.
13. Aydin, H., Aydın, G.Z.G. and Aydın, M.A., 2024. *The potential of light fidelity in smart home automation. Bulletin of Electrical Engineering and Informatics*, 13(5), pp.3155–3166.
14. FakhrHosseini, S., Lee, C., Lee, S.H. and Coughlin, J., 2024. A taxonomy of home automation: expert perspectives on the future of smarter homes. *Information Systems Frontiers*, pp. 1–18, doi:10.1007/s10796-024-10585-2.
15. Ansari, A.M., Nazir, M. and Mustafa, K., 2024. Ontology-based classification and detection of the smart home automation rules conflicts. *IEEE Access*, 12, pp. 85072–85088.

10 Vulnerability Analysis Using SQL Map in Smart Meters

Utkarsh Rajesh, Adarsh Mohapatra, Amit Hota, Bhupesh Singh Kushwaha, Mridulay Dixit, and O. V. Gnana Swathika

10.1 INTRODUCTION

In the dynamic landscape of smart metering within smart home networks, the burgeoning integration of these advanced systems has unraveled vulnerabilities susceptible to cyber threats [1]. This research navigates the intricate realm of smart meters, employing the SQL map penetration testing platform to meticulously simulate and expose potential risks [4]. Smart meters, pivotal in optimizing energy calculations and automating billing processes, have given rise to innovative solutions addressing inherent challenges [2]. Novel functionalities, such as embedded controllers and GSM modules, are explored to enhance communication and streamline bill generation [3]. Solutions to combat power theft leverage ZigBee for Internet of Things (IoT)–based connections and GSM modules for theft alerts, bolstering the security of Smart Grid infrastructure. Proposals extend beyond conventional power meters, incorporating cutting-edge technologies like NODEMCU with WiFi modules for secure data transmission to cloud servers and smart energy meters utilizing Fast Fourier transform to ensure precise energy calculations [3].

A significant focus is placed on the security risks associated with the rapid proliferation of smart devices in smart cities and grids, underscored by past incidents like the 2015 Ukraine power grid hack. While various smart metering system designs are scrutinized, the overarching theme emphasizes the imperative for comprehensive security measures [4]. This chapter introduces a compact smart meter design fortified with advanced security features, aiming to set a precedent for future developments. As the research unfolds, it calls for synchronization with power utility servers, integration with smart

DOI: 10.1201/9781003649700-10

homes, and rigorous penetration testing to fortify the system against potential cyber threats, ensuring the resilience of smart metering in the evolving landscape [5].

10.1.1 Securing the Future of Smart Grids: Addressing Vulnerabilities, Cyber Threats, and Innovation in Smart Metering Systems

The transition from analog to digital control in power systems, fueled by Smart Grid technology and the widespread adoption of smart meters, introduces both advantages and vulnerabilities. Adversaries exploit these vulnerabilities, causing substantial global economic losses nearing one hundred billion dollars annually due to electricity theft. This survey extensively examines electricity theft, exploring adversary tactics and existing detection methods, categorized into machine learning and measurement mismatch-based approaches [6].

This chapter also addresses the cybersecurity challenges faced by smart meters, emphasizing the imperative for information security in Smart Grid deployments [8]. Controlled laboratory experiments reveal the vulnerability of smart meters to common information security attacks, underscoring the need for robust countermeasures [7]. Additionally, the research delves into the security implications of Advanced Metering Infrastructure (AMI) in integrating renewable energy resources, stressing the necessity of robust security controls [8]. Within the context of smart home networks, the evolution of smart metering applications has unearthed vulnerabilities to cyber threats [9]. The utilization of the open-source penetration testing platform "SQL map" during simulation exposes potential risks in these advanced systems [9]. Smart meters seamlessly connect appliances, optimizing energy calculations and automating electricity bill estimates stored in a dedicated web portal.

In response to these challenges, innovative solutions have emerged. The integration of embedded controllers and GSM modules in smart meters introduces novel functionalities, enabling streamlined bill generation through SMS communication [10]. To combat power theft, ZigBee for IoT-based connections, coupled with GSM modules, is employed to provide theft alerts, bolstering the overall security of Smart Grid infrastructure [10]. Recognizing the limitations of traditional power meters, proposed solutions leverage cutting-edge technologies. The deployment of NODEMCU with a WiFi module ensures secure transmission of precalculated values to a cloud server, enhancing accessibility and data security. Similarly, a smart energy meter utilizes Fast Fourier transform to overcome non-linear loads, ensuring precise energy calculations and maintaining power quality [11]. In India, an Ethernet-based smart energy meter monitors power quality parameters and implements an automated system to manage loads, ultimately reducing electricity costs [11].

The discussion expands to smart cities and grids, emphasizing the security risks associated with the rapid adoption of smart devices. Instances like the 2015 Ukraine power grid hack underscore the need for stringent security measures in Smart Grid designs [12]. While various smart metering system designs are explored, the overarching theme emphasizes the necessity of comprehensive security measures.

10.1.2 Integration of Intelligent Power Metering Systems for Enhanced Residential Energy Management

In today's homes, a big change is happening in how we use energy, thanks to smart power meters. These advanced systems are transforming the way we manage and make the most of energy at home.

The study [12] introduces an innovative and efficient way of creating a smart meter. This smart meter was successfully integrated using an embedded controller and GSM module with the capability of communication through SMS. The smart meter can automatically estimate the predefined values of per kilowatt energy consumption and generate the bill to the user via SMS through mobile networks.

Power theft cases are usually common in recent days which creates an imbalance in electricity bill transactions. To minimize these issues, a smart energy meter approach is created. The proposed system [13] consists of an energy meter chip that is capable of automatically calculating the electricity bill displayed on the LCD from the user's side. An IoT-based connection is established using ZigBee protocol between the user's side and the substation. This system also utilizes a GSM module responsible for sending alerts in case of power theft to the user as well as the substation.

Traditional power meters lack full-duplex communication support that is one of the major drawbacks. To solve this problem, Barman et al. [14] have proposed a system that utilized NODEMCU with a WiFi module for securely transmitting the precalculated values of electricity bill to a cloud server. This improves the consumer/producer side accessibility of the data enabling a secure analysis, control, and consumption of energy.

The paper [15] focuses on reliable connectivity over the last mile scenarios in IoT applications and challenges related to implementing long-range connectivity to individually connected smart meters in the home area network (HAN). The authors highlight the emergence of low-power wide area network (LPWAN) technologies as a capable solution for long-range connectivity in IoT networks as they provide low-cost, reliable, low-power, long-range last mile technology for smart meters. A further prototype is demonstrated using LoRa connectivity solutions for smart metering applications in a diverse and challenging environment.

The traditional energy meters are equipped with non-linear loads and are difficult to analyze and monitor the energy consumption status. This leads to distortion of current and voltage signals, further affecting the quality of power. To overcome this glitch [16], the proposed smart energy meter uses Fast Fourier transform to calculate the energy. The voltage and current values are calculated according to the various power calculation techniques by the microcontroller and verified using power quality analyzers for successfully monitoring in a "thingspeak" cloud platform.

In a diverse country like India, there is still no clarity observed on power quality monitoring due to the varying power across the state. The power meters also lack a user-friendly interface for displaying the energy consumption data. The paper [17] implements an approach to monitor power quality parameters via an Ethernet-based smart energy meter. The energy consumption data is stored in a server for further processing. A simulation tool "LABVIEW" is utilized to calculate and monitor the status of power quality measurements. On the household level, the energy meter identifies the appliances as inductive, capacitive, and resistive. An automated system is developed that manages the loads to cycle reactive power. This helps in reducing electricity bill costs and improving the power factor [17].

10.2 METHODOLOGY

Smart metering applications reveal cyber vulnerabilities, as demonstrated. Metering devices connect appliances to a common terminal in a smart home network. These advanced devices provide efficient calculations of energy units per kilowatt hour, and can automatically estimate the electricity bill based on electricity rates. A separate database stores each user's monthly electricity estimates in a web portal.

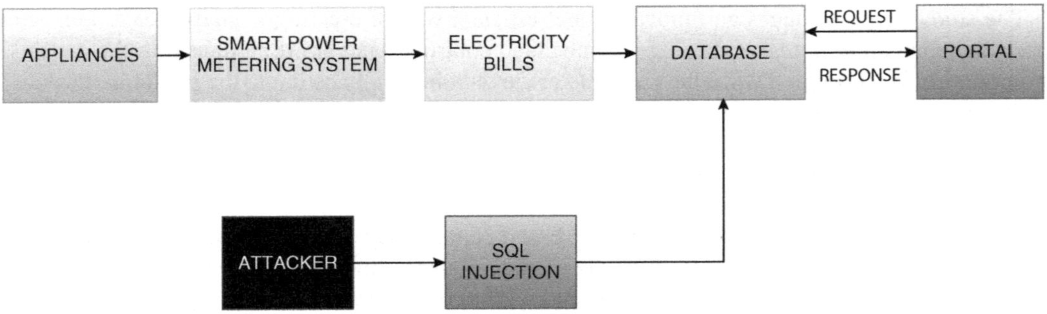

FIGURE 10.1 Block diagram of SQL injection.

To perform a simulation, an open-source platform named "SQLmap" is used as in Figure 10.1. This platform provides a penetration testing tool that helps in exploiting the features of SQL injection. The user's database is tested using various types of SQL injection vulnerabilities during the reconnaissance.

10.2.1 Algorithmic Insights

- **Preparing the Environment**:
  - The code starts by setting up some basic things for the program.
  - It tells the system not to create extra files when the program runs.
- **Importing Tools**:
  - It brings in some tools that the program will use, both standard ones and some custom ones from other parts of the program.
- **Getting Ready**:
  - It does some setup work, like making sure everything is in the right place and fixing some things that might cause issues later.
- **Logging In**:
  - It decides how much information the program should share with us while it's running, like if it should tell us every little detail or just the important stuff.
- **Figuring Out Where It Is**:
  - It figures out where it's located on the computer. This is like the program finding its own home.
- **Listening for Commands:**
  - It checks what commands you gave it when you ran the program. For example, did you ask it to be a server or a client?
- **Doing the Main Job:**
  - Based on your commands, it does the main tasks of the program. If you asked it to be a server, it sets up a service to listen for requests. If you asked it to be a client, it connects to a service.
- **Telling You What's Happening:**
  - While it's working, it might share some messages with you, especially if something interesting or important is happening.
- **Checking If It's the Boss:**
  - It checks if it's the main program that's supposed to run everything. If yes, it starts the main tasks.

10.3 RESULT AND DISCUSSION

In our investigation of smart metering applications, vulnerability analysis using the SQL injection tool "SQL map" revealed two critical attack types as in Figure 10.2. A blind Boolean-based attack relies on predicting the contents of the database through Boolean responses, while a Union Query-based attack combines embedded and original statements to extract data. The exploratory phase was focused on understanding database structures, identifying tables, and enumerating columns.

Implications include the potential for unauthorized access and manipulation of power consumption data, posing a risk to the integrity of user information. Mitigation strategies include secure coding, regular security audits, and user education. The urgency to address these vulnerabilities is highlighted to ensure the integrity of energy consumption data, underscoring the need for robust security measures in smart metering systems. Future research should prioritize the development of effective security measures against evolving cyber threats.

When the simulation is performed, it generates a request from the server to the client for sharing the data from the specified portal and displays the log status of possible injectable vulnerability points. It then classifies these vulnerabilities into different classes of SQL injection-based attacks supported by the open-source platform "SQL map." In our results, two distinct types of attacks are initiated which are as follows:

```
[00:11:12] [INFO] GET parameter 'id' is 'Generic UNION query (NULL) - 1 to 20 columns' injectable
GET parameter 'id' is vulnerable. Do you want to keep testing the others (if any)? [y/N]
sqlmap identified the following injection points with a total of 13 HTTP(s) requests:
...
Place: GET
Parameter: id
    Type: boolean-based blind
    Title: AND boolean-based blind - WHERE or HAVING clause
    Payload: id=1 AND 6486=6486

    Type: UNION query
    Title: Generic UNION query (NULL) - 3 columns
    Payload: id=1 UNION ALL SELECT ':xjv:'||'WSZGEAHHop'||':ovt:', NULL, NULL··
...
[00:11:15] [INFO] testing SQLite
[00:11:15] [INFO] confirming SQLite
[00:11:15] [INFO] actively fingerprinting SQLite
[00:11:15] [INFO] the back-end DBMS is SQLite

Web application technology: PHP 5.3.3, Apache 2.2.16
back-end DBMS: SQLite
[00:11:15] [INFO] fetching columns for table 'users' in database 'SQLite_masterdb'
Database: SQLite_masterdb
Table: users
[3 columns]
+-------------+-------------+
| Column      | Type        |
+-------------+-------------+
| id          | INTEGER     |
| name        | TEXT        |
| surname     | TEXT        |
+-------------+-------------+
```

FIGURE 10.2 Outcome after SQL injection.

- **Boolean-based Blind**: It is a type of SQL-based injection where the attacker predicts the database content according to the Boolean value responses from the application. The attacker injects an SQL code into the input environment. The injected program instead of directly accessing the data asks true or false questions regarding the database content. A page response by the portal is displayed based on the Boolean nature of the injected statement.
- **Union Query-based**; It is another type of SQL injection where the attacker uses a set theory operator known as UNION in SQL that combines the injected and original statements executed by the application through UNION ALL SELECT, enabling the attacker to extract the data.
- During the reconnaissance phase of a cyber-attack, the attacker collects information about the structure of the database, including its tables and columns. These columns act as specific attributes and identifiers of the relational database, storing various data types. The attacker aims to enumerate the database table's columns, retrieving their characteristics and functionalities, before exploiting any vulnerabilities in the database's structure.

10.4 CONCLUSION

In conclusion, this research illuminates the dual nature of smart metering applications—offering transformative benefits to energy management while revealing significant cybersecurity vulnerabilities. The simulation using "SQL map" underscores the tangible threats posed by SQL injection-based attacks, specifically Boolean-based blind and Union Query-based, emphasizing the imperative for robust cybersecurity measures in securing smart metering infrastructures.

The exploration of innovative solutions in related studies, ranging from IoT-based connections to advanced communication protocols like GSM, ZigBee, and LoRa, demonstrates a commitment to addressing challenges such as power theft and communication limitations. However, the conclusion underscores that, despite these advancements, a proactive and comprehensive cybersecurity approach is paramount. As smart meters become increasingly integral to modern energy systems, continuous monitoring, encryption, and secure communication protocols are crucial for ensuring the reliability and integrity of these networks. The research calls for a delicate balance between technological innovation and stringent cybersecurity measures. The evolving landscape demands a dynamic and adaptive approach to safeguarding smart energy systems. As we navigate the dynamic technological advancements, a resilient and secure foundation is essential for the sustainable evolution of smart metering applications in the pursuit of efficient and secure energy management.

REFERENCES

1. Chang, Y.H., Jirutitijaroen, P., and, Ten, C.W. 2010. A simulation model of cyber threats for energy metering devices in a secondary distribution network. In: *2010 5th International Conference on Critical Infrastructure (CRIS)*, Beijing, China (pp. 1–7). IEEE.
2. Kumar, H., Alvarez, O.A., and, Kumar, S. 2023. Experimental evaluation of smart electric meters resilience under cyber security attacks. *IEEE Access*, 11. doi:10.1109/ACCESS.2023.
3. Ur-Rehman, O., Zivic, N., and, Ruland, C. 2015. Security issues in smart metering systems. In: *2015 IEEE International Conference on Smart Energy Grid Engineering (SEGE)*, Oshawa, ON, Canada (pp. 1–7). IEEE.

4. Win, L.L., and, Tonyalı, S. 2021. Security and privacy challenges, solutions, and open issues in smart metering: A review. In: *2021 6th International Conference on Computer Science and Engineering (UBMK)*, Ankara, Turkey (pp. 800–805). IEEE.
5. Hafiz Abdullah, M.D., Hanapi, Z.M., Zukarnain, Z.A., and, Mohamed, M.A. 2015. Attacks, vulnerabilities, and security requirements in smart metering networks. *KSII Transactions on Internet and Information Systems (TIIS)*, 9(4), pp. 1493–1515.
6. Xia, X., Xiao, Y., Liang, W., and, Cui, J. 2022. Detection methods in smart meters for electricity thefts: A survey. *Proceedings of the IEEE*, 110(2), pp. 273–319.
7. Shuaib, K., Trabelsi, Z., Abed-Hafez, M., Gaouda, A., and, Alahmad, M. 2015. Resiliency of smart power meters to common security attacks. *Procedia Computer Science*, 52, pp.145–152.
8. Menon, V.D., Kumar, J.T., Sabhanayagan, M., Ramkumar, A., and, Rajesh, K. 2019. Cyber security for smart meters. In: *2019 IEEE International Conference on Intelligent Techniques in Control, Optimization and Signal Processing (INCOS)*, Tamil Nadu, India (pp. 1–5). IEEE.
9. Khattak, A.M., Khanji, S.I., and, Khan, W.A. 2019. Smart meter security: Vulnerabilities, threat impacts, and countermeasures. In: *Proceedings of the 13th International Conference on Ubiquitous Information Management and Communication (IMCOM) 2019*, Phuket, Thailand (pp. 554–562). Springer International Publishing.
10. Saha, S., Mondal, S., Saha, A., and, Purkait, P. 2018. "Design and implementation of IoT based smart energy meter." In: *2018 IEEE Applied Signal Processing Conference (ASPCON)*, Kolkata, India (pp. 19–23). IEEE.
11. Hasan, M., Iqbal, F., C.K. Patrick, Hung, B.C.M., and, Rafferty, L. 2019. IoT security and privacy: Challenges and solutions. *IEEE Consumer Electronics Magazine*, 8(2), 37–43.
12. Rahman, M.M., Islam, M.O., and, Salakin, M.S. 2015. May. Arduino and GSM based smart energy meter for advanced metering and billing system. In: *2015 International Conference on Electrical Engineering and Information Communication Technology (ICEEICT)*, Dhaka, Bangladesh (pp. 1–6). IEEE.
13. Preethi, V., and Harish, G. 2016. Design and implementation of smart energy meter. In: *2016 International Conference on Inventive Computation Technologies (ICICT)*, Coimbatore, India (Vol. 1, pp. 1–5). IEEE.
14. Barman, B.K., Yadav, S.N., Kumar, S., and, Gope, S. 2018. IOT based smart energy meter for efficient energy utilization in smart grid. In: *2018 2nd International Conference on Power, Energy, and Environment: Towards Smart Technology (ICEPE)*, Shillong, India (pp. 1–5). IEEE.
15. Jain, S., Pradish, M., Paventhan, A., Saravanan, M., and, Das, A. 2018. Smart energy metering using LPWAN IoT technology. In: *ISGW 2017: Compendium of Technical Papers: 3rd International Conference and Exhibition on Smart Grids and Smart Cities,* New Delhi, India (pp. 19–28). Springer Singapore.
16. Kumar, L.A., Indragandhi, V., Selvamathi, R., Vijayakumar, V., Ravi, L., and, Subramaniyaswamy, V. 2021. Design, power quality analysis, and implementation of smart energy meter using internet of things. *Computers & Electrical Engineering*, 93, p.107203.
17. Das, H., and, Saikia, L.C. 2017. Ethernet based smart energy meter for power quality monitoring and enhancement. In: *2017 Recent Developments in Control, Automation & Power Engineering (RDCAPE)*, Noida, India (pp. 187–191). IEEE.

Index

For Product Safety Concerns and Information please contact our EU representative GPSR@taylorandfrancis.com Taylor & Francis Verlag GmbH, Kaufingerstraße 24, 80331 München, Germany

Batch number: 10767716

Printed by Printforce, the Netherlands